Characterization, Modeling, Monitoring, and Remediation of FRACTURED ROCK

Committee on Subsurface Characterization, Modeling, Monitoring, and Remediation of Fractured Rock

Committee on Geological and Geotechnical Engineering

Board on Earth Sciences and Resources

Division on Earth and Life Studies

A Consensus Study Report of
The National Academies of
SCIENCES · ENGINEERING · MEDICINE

THE NATIONAL ACADEMIES PRESS
Washington, DC
www.nap.edu

THE NATIONAL ACADEMIES PRESS **500 Fifth Street, NW** **Washington, DC 20001**

This activity was supported by the National Aeronautics and Space Administration under Grant No. NNX12AR45G, the U.S. Nuclear Regulatory Commission under Grant No. NRC-04-09-167, and the U.S. Department of Energy under Contract No. DE-PI000010/DE-DT0002974. Any opinions, findings, conclusions, or recommendations expressed in this publication do not necessarily reflect the views of any organization or agency that provided support for the project.

International Standard Book Number-13: 978-0-309-37372-2
International Standard Book Number-10: 0-309-37372-7
Digital Object Identifier: https://doi.org/10.17226/21742

Additional copies of this publication are available from the National Academies Press, 500 Fifth Street, NW, Keck 360, Washington, DC 20001; (800) 624-6242 or (202) 334-3313; http://www.nap.edu.

Cover design by Eric Edkin.

Printed in the United States of America

Suggested citation: National Academies of Sciences, Engineering, and Medicine. 2020. *Characterization, Modeling, Monitoring, and Remediation of Fractured Rock*. Washington, DC: The National Academies Press. https://doi.org/10.17226/21742.

The National Academies of

SCIENCES • ENGINEERING • MEDICINE

The **National Academy of Sciences** was established in 1863 by an Act of Congress, signed by President Lincoln, as a private, nongovernmental institution to advise the nation on issues related to science and technology. Members are elected by their peers for outstanding contributions to research. Dr. Marcia McNutt is president.

The **National Academy of Engineering** was established in 1964 under the charter of the National Academy of Sciences to bring the practices of engineering to advising the nation. Members are elected by their peers for extraordinary contributions to engineering. Dr. John L. Anderson is president.

The **National Academy of Medicine** (formerly the Institute of Medicine) was established in 1970 under the charter of the National Academy of Sciences to advise the nation on medical and health issues. Members are elected by their peers for distinguished contributions to medicine and health. Dr. Victor J. Dzau is president.

The three Academies work together as the **National Academies of Sciences, Engineering, and Medicine** to provide independent, objective analysis and advice to the nation and conduct other activities to solve complex problems and inform public policy decisions. The National Academies also encourage education and research, recognize outstanding contributions to knowledge, and increase public understanding in matters of science, engineering, and medicine.

Learn more about the National Academies of Sciences, Engineering, and Medicine at **www.nationalacademies.org**.

The National Academies of
SCIENCES · ENGINEERING · MEDICINE

Consensus Study Reports published by the National Academies of Sciences, Engineering, and Medicine document the evidence-based consensus on the study's statement of task by an authoring committee of experts. Reports typically include findings, conclusions, and recommendations based on information gathered by the committee and the committee's deliberations. Each report has been subjected to a rigorous and independent peer-review process and it represents the position of the National Academies on the statement of task.

Proceedings published by the National Academies of Sciences, Engineering, and Medicine chronicle the presentations and discussions at a workshop, symposium, or other event convened by the National Academies. The statements and opinions contained in proceedings are those of the participants and are not endorsed by other participants, the planning committee, or the National Academies.

For information about other products and activities of the National Academies, please visit www.nationalacademies.org/about/whatwedo.

COMMITTEE ON SUBSURFACE CHARACTERIZATION, MODELING, MONITORING, AND REMEDIATION OF FRACTURED ROCKS[1]

DAVID E. DANIEL (NAE) (*Chair*), The University of Texas System
LISA ALVAREZ-COHEN (NAE), University of California, Berkeley
WILLIAM DERSHOWITZ, Golder Associates Inc., Redmond, Washington
HERBERT H. EINSTEIN, Massachusetts Institute of Technology, Cambridge
CARL GABLE, Los Alamos National Laboratory, New Mexico
FRANKLIN M. ORR, JR., Stanford University, California (*resigned December 2014*)
DAVID REYNOLDS, Geosyntec Consultants, Inc., Kingston, Ontario, Canada
J. CARLOS SANTAMARINA (NAE), Georgia Institute of Technology, Atlanta
ALLEN M. SHAPIRO, U.S. Geological Survey, Reston, Virginia
KAMINI SINGHA, Colorado School of Mines, Golden

Staff

SAMMANTHA L. MAGSINO, Study Director
ERIC J. EDKIN, Senior Program Assistant

[1] Staff and affiliations current as of September 2015.

COMMITTEE ON GEOLOGICAL AND GEOTECHNICAL ENGINEERING[1]

PATRICIA CULLIGAN (*Chair*), Columbia University, New York, New York
JEAN-LOUIS BRIAUD, Texas A&M University, College Station
JOHN R. CRAYNON, Virginia Polytechnic Institute and State University, Blacksburg
MURRAY W. HITZMAN, Colorado School of Mines, Golden
MARTIN W. McCANN, Jack R. Benjamin and Associates, Inc., Menlo Park, California
PRISCILLA NELSON, Colorado School of Mines, Golden
JUAN M. PESTANA, University of California, Berkeley
JAMES R. RICE, Harvard University, Cambridge, Massachusetts
GLENN RIX, Geosyntec Consultants, Inc., Kennesaw, Georgia

Staff

SAMMANTHA L. MAGSINO, Senior Program Officer
COURTNEY R. DEVANE, Administrative Coordinator

[1] Staff and affiliations current as of September 2015.

BOARD ON EARTH SCIENCES AND RESOURCES[1]

[1] Staff and affiliations current as of September 2015.

Reviewers

This Consensus Study Report was reviewed in draft form by individuals chosen for their diverse perspectives and technical expertise. The purpose of this independent review is to provide candid and critical comments that will assist the National Academies of Sciences, Engineering, and Medicine in making each published report as sound as possible and to ensure that it meets the institutional standards for quality, objectivity, evidence, and responsiveness to the study charge. The review comments and draft manuscript remain confidential to protect the integrity of the deliberative process.

We thank the following individuals for their review of this report:

MICHAEL ANNABLE, University of Florida
PETER COOK, CSIRO Land and Water
RUSSELL DETWILER, University of California, Irvine
MARK HARKNESS, GE Research
WALTER ILLMAN, University of Waterloo
LAURA PYRAK-NOLTE, Purdue University
KENNETH RAVEN, Geofirma Engineering Ltd.
NICHOLAS SITAR, University of California, Berkeley
JESSE YOW, Lawrence Livermore National Laboratory

Although the reviewers listed above provided many constructive comments and suggestions, they were not asked to endorse the conclusions or recommendations of this report nor did they see the final draft before its release. The review of this report was overseen by **George Hornberger,** Vanderbilt University. He was responsible for making certain that an independent examination of this report was carried out in accordance with the standards of the National Academies and that all review comments were carefully considered. Responsibility for the final content rests entirely with the authoring committee and the National Academies.

Contents

SUMMARY **1**

1 INTRODUCTION **9**
The Context, 9
The Committee's Task, 10
Previous National Academies Studies, 12
The Heterogeneity of Fractured Rock, 13
Contaminants in Fractured Rock, 15
Fate and Transport, 17
Knowledge from Related Areas, 18
Report Organization, 18

2 PHYSICAL CHARACTERISTICS OF FRACTURED ROCK CONTROLLING FLOW AND TRANSPORT **20**
Qualitative Fracture Description, 20
Quantitative Fracture Description, 22
Importance of Fracture Genesis, 24

3 FLOW AND TRANSPORT: UNDERLYING PROCESSES **26**
Advection and Dispersion, 26
Hydro-Mechanical Coupling, 29
Chemical Processes: Diffusion and Reaction, 29
Biological Processes, 34
Mixed-Fluid Conditions, 36
Fines Migration and Entrapment: Emergent Transport Processes, 41
Heat Transport and Thermal Processes, 43

4 CONCEPTUAL AND NUMERICAL MODEL DEVELOPMENT **46**
Defining and Developing Hydrostructural Models, 47
Quantifying the Hydrostructural Model, 53
Scoping Calculations to Assess and Refine Models and Uncertainties, 57
Analysis Tools to Inform Modeling, 58
Types of Numerical Models, 61
Upscaling and Model Simplification, 63
Numerical Model Analysis, 64
Analysis and Resources, 67

5 METHODS FOR SITE CHARACTERIZATION AND MONITORING **68**
Geomechanical Characterization, 68
Geometric Characterization, 70
Hydraulic Characterization of Fractured Rock, 71
Geophysical Characterization of Fractured Rock, 82
Geochemical Characterization of Fractured Rock Systems, 86
Biological Characterization of Fractured Rock, 91

6 **REMEDIATION OF FRACTURED ROCK** **94**
Difficulties of Remediation in Fractured Rock, 95
Potential Technologies to Remediate Organic Compounds, 99
Potentially Applicable Approaches for Radionuclides, 104
Natural Attenuation, 105
Important Considerations in Fractured Rock Remediation, 105
The Path Forward, 110

7 **DECISION MAKING** **112**
Decision Making and Updating in the Context of this Report, 112
Observational Method in Geotechnical Engineering, 114
Evolving the Observational Method to Geoenvironmental Engineering, 116

8 **SYNTHESIS OF RECOMMENDATIONS** **119**
An Integrated Approach, 119
Interactions Between Rock Matrix and Fractures, 120
Processes and Coupled Processes, 121
Characterization Techniques and Tools, 123
Modeling, 124
Remediation and Monitoring, 125
The Observational Approach, 127
Final Thoughts, 128

REFERENCES **129**

APPENDIXES
A COMMITTEE MEMBER BIOGRAPHIES **157**
B OPEN SESSION AGENDAS **161**

Summary

Fractured rock is the host or foundation for innumerable engineered structures related to energy, water, waste, and transportation. Characterizing, modeling, and monitoring fractured rock sites is critical to the functioning of those infrastructures, as well as to optimizing resource recovery and contaminant management. Fractured rock is defined in this report as a mass of rock matrix broken up by fractures. The rock may be crystalline with nominal porosity (i.e., many igneous rocks) or granular with varying amounts of cementation or porosity (i.e., sedimentary rocks).

At the request of the National Aeronautics and Space Administration, the U.S. Nuclear Regulatory Commission, and the U.S. Department of Energy, the National Academies of Sciences, Engineering, and Medicine (the National Academies) conducted a study to address issues relevant to subsurface fluid flow and transport in fractured rock systems. The National Academies convened an expert committee of researchers and practitioners to examine the state of the practice and state of the art in characterization of fractured rock and of the mechanical, chemical, and biological processes related to subsurface contaminant fate and transport. The committee also considered conceptual modeling of fractured rock and fluid and contaminant transport within it, the detection and monitoring of fluid and contaminant pathways and travel times, and remediation of contaminated sites in the event of system failures. The committee's Statement of Task is in Box S.1. The committee interpreted its task to include consideration of all types of naturally fractured rock systems beneath the vadose zone to depths of 3 to 5 kilometers.

This report examines new developments, knowledge, and approaches to engineering at fractured rock sites since publication of the 1996 National Research Council report *Rock Fractures and Fluid Flow: Contemporary Understanding and Applications*. Fundamental understanding of the physical nature of fractured rock has changed little since 1996, but many new characterization tools have been developed, and there is now greater appreciation for the importance of chemical and biological processes that can occur in the fractured rock environment. Findings in this report can be applied to all types of engineered infrastructure and engineered in situ processes, but especially to engineered repositories for buried or stored waste and to fractured rock sites that have been contaminated as a result of past disposal or other practices. Impacts from artificially induced fractures to enhance hydrocarbon recovery (i.e., hydraulic fracturing) are not part of this report.

This report describes how existing tools—some only recently developed—can be used to increase the accuracy and reliability of engineering design and management given the interacting forces of nature. Examples of science and engineering research that can advance those tools or develop new tools are provided. Recommendations are organized by theme and include recommendations for practice and research in a given thematic area. The recommendations are high level and intended to help the practitioner, researcher, and decision maker take a more interdisciplinary approach to engineering in the fractured rock environment. An integrated systems approach for engineering fractured rock sites is emphasized, and recommendations are presented in a broader systems context. The recommendations are not intended for specific groups of experts or professional sectors. Using interdisciplinary approaches makes it possible to conceptualize and model the fractured rock environment with acceptable levels of uncertainty and reliability. They make it possible to design systems that maximize remediation effectiveness and long-term performance. The fractured rock environment can be complex, but it is subject to known laws of nature.

CONTAMINANTS IN THE SUBSURFACE

Groundwater and contaminant transport and storage take place within void spaces (i.e., porosity) in both the rock matrix and fractures. However, although flow is dominated generally by only a few of the rock fractures in a fractured rock system, storage takes place predominantly within the rock matrix. This fundamental distinction between transport and storage has profound implications to fractured rock characterization, modeling, monitoring, and remediation.

Contaminant fate and transport in the fractured rock environment are affected significantly by contaminant solubility. Minimally water-soluble contaminants such as non-aqueous phase liquids can flow at rates and directions different from those of groundwater. Those that are denser than water (dense non-aqueous phase liquids such as trichloroethene) migrate downward in fractured rock until a barrier is encountered, making them difficult to locate and characterize. Water soluble contaminants can travel with water great distances from their sources, react with surrounding geologic and organic materials, or precipitate from solution.

BOX S.1 Statement of Task

Geological and geotechnical characterizing, modeling, and monitoring of the subsurface are integral to safe, economical, and responsible development, maintenance, operation, remediation, and decommissioning of infrastructure related to energy, water, waste, and transportation. Modeling and monitoring fluid travel paths and velocities through subsurface fractures and matrix are among the most significant engineering challenges associated with these tasks. An ad hoc committee of the National Academies of Sciences, Engineering, and Medicine will conduct a study to address issues relevant to subsurface flow and contaminant transport in fractured media, including low permeability and low porosity media, as well as in deep (3 to 5 kilometers) fracture systems. Subsurface characterization, modeling, monitoring, and remediation (SCMMR) issues applicable throughout the lifecycle of engineered facilities that have the potential to release contaminants and pose risk to groundwater quality will be considered. As part of its information gathering, the committee will convene a workshop to examine the state of art and state of practice in

- Subsurface fracture and matrix characterization, especially relevant geotechnical, hydrological, and geochemical properties, and the development of conceptual models;
- Detection of fluid and contaminant pathways and travel times;
- Detection and modeling of factors that affect changes in geotechnical and hydrological properties over time (e.g., decades to millennia), including thermal, hydrological, chemical, and mechanical processes;
- Groundwater and contaminant transport modeling, monitoring, and remediation, and how these can aid decision making during facility design, operation, remediation, and decommissioning;
- Early indicators (e.g., change in fracture properties, moisture levels, background chemistry) of system failures resulting in unintentional release of fluids; and
- Potential mitigation measures to eliminate or reduce adverse impacts of system failures and related releases to the environment.

SCMMR knowledge generated from energy industry practice as well as carbon sequestration investigations will be considered. The committee will issue a final report that will include findings and conclusions with respect to (i) where research and development could improve the current state of art in SCMMR, (ii) where incorporation of scientific and technical advances could enhance the state of practice in SCMMR, and (iii) where enhanced science-based understanding could inform federal regulations, policies, and implementing guidance.

UNDERSTANDING THE FRACTURED ROCK ENVIRONMENT

Adequate characterization and modeling of the fractured rock environment and the potential for contaminant fate and transport is critical when siting, designing, and managing infrastructure that could release contaminants. Modeling is also necessary to assess and design remediation schemes for areas where contaminants have already entered the subsurface. First steps in characterization include understanding the geologic setting and history so that the genesis of fractures can be understood and the most important hydrogeologic features can be defined. Expertise from several fields is best integrated at the earliest stages of project development and during data collection, analyses, and decision making to account for fractured rock processes that affect contaminant transport, storage, and transfer between the rock matrix and fractures. Such an integrated approach may yield better recognition of, for example, the ways discrete fractures may dominate flow and transport, or the reliability of measurements across large areas from long, open boreholes.

Recommendation 1. Take an interdisciplinary approach to engineering in fractured rock and use site geologic, geophysical, geomechanical, hydrologic, and biogeochemical information to conceptualize transport pathways, storage porosities, fate-and-transport mechanisms, and the coupled processes that control rock fracture-matrix interactions.

Use available geologic, hydrogeologic, and geophysical information to conceptualize flow pathways, fluid and contaminant storage, alteration, or attenuation through geochemical reactions and transformations via biological processes. When there are little data, for example, during early stages of characterization, fractured rock hydrogeology can be conceptualized by first developing a site conceptual model that assumes fluid storage occurs primarily in the rock matrix, and flow occurs primarily within fractures (i.e., a dual porosity/single permeability environment). Simplify the conceptual model (i.e., to single effective porosity and permeability) or add further detail (i.e., to dual porosity/dual permeability or discrete fracture networks) only when justified by site-specific evidence. As flow and transport are evaluated, recognize the limitations of advection-dispersion-diffusion approaches to solute transport solutions, particularly with regard to scale. Parameters developed at the meter scale, for example, may not be applicable at the kilometer scale.

Recommendation 2. Estimate the potential for contaminant to be transported into, stored in, and transported back out of rock matrix over time.

Fractures and fracture networks can be discrete flow pathways, barriers, or provide void space for storage. Understanding fracture geometries, possible heterogeneities, and interactions between the rock matrix and fractures (e.g., advection, diffusion, sorption, biodegradation, filtration, and capillary processes) is critical to understanding fate and transport. The storage of contaminants in the rock matrix and their slow diffusive transfer from the matrix to fluid flowing in adjacent fractures can result in remediation time frames of decades or centuries. Failure to quantify and differentiate between stored and mobile contaminants and site-specific storage and release mechanisms can lead to gross miscalculation of subsurface contaminant distribution and the effectiveness of remedial measures (e.g., pump and treat). The effects of rock fracture and matrix void sizes on capillary and wetting processes, bioactivity, abiotic reactions, and multiphase flow should also be evaluated.

Recommendation 3. Characterize chemical, biological, thermal, mechanical, and hydraulic processes, their interactions, and the conditions that can lead to their coupling to better understand transport through discrete fractures and contaminant transfer between fractures and rock matrix.

The importance of coupled processes in fate and transport is not always recognized or addressed. Individual and combined processes can change how fluids containing contaminants, migrating fines, and colloids interact with the rock matrix and fracture infillings and coatings. It is necessary to take into account the possibility of flow localization within fractures, channels within fracture planes and intersections, and hydro-mechanical coupling that results in changes in fracture and channel geometries. The effects of these changes on advection, dispersion, and diffusion in the interpretation and prediction of transport rates need to be understood to characterize fate and transport. There is also the potential for mineral fines to clog pathways or actually enable contaminant transport (i.e., Pickering emulsions). Fluid properties (e.g., miscibility and mutual solubility, density, viscosity, and acidity) and reactivity with rock or infilling material can also be important. Immiscible fluids exist as a separate phase from water and may impede or block groundwater flow paths, forcing groundwater to flow around them. When there is the possibility of thermal changes, assess subsurface heat conduction and temperature changes under static- and advective-flow conditions, including occurrences of flow localization. It is also important to consider how all of these factors—particularly stress, temperature, and groundwater chemistry—change with depth at a site.

Characterization of microbial communities and activities in fractured rock allows better characterization and prediction of fluid and contaminant fate and transport, as well as more effective application of bioremediation technologies. Subsurface microbial communities can affect physical and geochemical characteristics and may be responsible for a variety of dynamic processes including mineral formation and dissolution, as well as changes in redox chemistry, fluid surface tension, and acidity. There are few comprehensive applications to understand microbial community structure and function in fractured rock environments, and, given their importance in the geochemical properties and potential for bioremediation in fractured rock, more research in this area represents a singularly important investment.

Recommendation 4. Expand research to define and quantify microbial influences on fluid and contaminant fate and transport in fractured rock over timescales relevant to contaminant remediation processes.

Reliable, reproducible, quantitative, and statistically valid experimental information on the spatial and temporal dynamics of biological communities in fractured rock is needed to understand the processes that take place, their effects, and their rates of occurrence. Knowledge of biological processes in fractured rock has advanced in the past decade because of improvements in molecular tools (e.g., genome sequencing at the individual cell level is now possible), but the study of in situ biological systems, especially for rock at great depth, is still impractical, and cultivating microorganisms from fractured rock environments remains difficult. However, by coupling advanced molecular tools with computational modeling, it should now be possible to systematically address fundamental questions.

Several areas of research could be particularly useful, such as identification of parameters that determine the extents of phylogenetic, genetic, and functional diversity of microbial communities in fractured rock environments; the influence of microbial populations on hydraulic and hydrogeochemical characteristics; the changes of phylogenetic and functional structures in microbial communities across various spatial and temporal scales (including associated relationships with hydrogeochemistry and the anoxic, high-salinity, high-temperature, and high-stress conditions of the deep environment); and the microbial community response to environmental perturbations in fractured rock environments.

CHARACTERIZATION TECHNIQUES AND TOOLS

Site characterization and monitoring tools and techniques for geologic, geophysical, geochemical, hydraulic, biological, and geomechanical observation, testing, and sampling have improved considerably since 1996. For example, remote sensing tools such as light detection and ranging permit large-scale fracture mapping at a high resolution, and borehole geophysical and imaging methods, pumping tests, and flow meter measurements help to identify zones and rates of flow. Advanced numerical algorithms now allow more information to be extracted from these and other test data, essentially creating opportunities for three-dimensional tomographic representations.

Most of these methods, however, are limited in their ability to allow the mapping of fracture characteristics at the needed scales and resolution. Borehole sampling and testing, for example, can be costly and provide limited information on a relatively small scale. Compressive stresses can complicate their use at the kilometer-scale depth. Cross-hole tracer or other hydraulic testing used to understand larger-scale issues such as fracture connectivity and flow dimension require multiple potentially costly boreholes. Remote methods such as seismic imaging provide only the means to infer rather than observe subsurface characteristics. Cross-hole radar and micro-seismic tomography are well suited for sites at depths, within their respective ranges.

Recommendation 5. Improve characterization and monitoring through new and expanded research in surface- and borehole-based geologic, geophysical, geochemical, hydraulic, biologic, and geomechanical technologies.

There is a need to advance research on technologies that use in situ measurements (e.g., innovative tracer tests and flexible liners) to characterize and monitor the volumes, spatial distribution, and transfer of contaminants between rock matrix and fractures. Fracture mapping techniques (e.g., thermal, electrical resistivity, and geochemical) could be improved to identify conductivity and flow in fractures, calculate in situ gradients, and determine flow directions. Geophysical techniques such as micro-seismic, electrical, nuclear magnetic resonance, and radar could be advanced to track tracers or contaminants remotely, including between injection and withdrawal locations. Research is needed on how to extend use of geophysical tools commonly used in porous media to fractured rock applications. Techniques such as seismic, microseismic, and hydraulic tomography, hydraulic interference, and tracer testing techniques need to be developed that allow for characterization of flow paths at different scales, and that advance joint inversion methodologies for fracture process parameterization. An important advance in characterization would be to determine ways that geophysical responses to fracture locations and geometries could be incorporated into fluid flow models.

CONCEPTUAL MODELING

Without a conceptual framework, numerical modeling will likely misrepresent hydrologic behavior at even the smallest project site. Conceptual model templates are available for different geologic environments and can be used to develop suitable conceptual models that incorporate the rock and hydrologic structures in a systematic way. Such conceptual models are called hydrostructural models. This kind of conceptual modeling promotes interdisciplinary approaches to decision making.

Because fractures can occur at the micron to kilometer scale, the full range of fluid fate and transport possibilities needs to be considered to determine which fractures and physical characteristics—and at what scales—are of hydrologic importance. However, current numerical modeling practice often relies on single-porosity, equivalent-continuum numerical models that are unable to account for discrete fracture flow pathways and fracture-matrix interactions more suitable to a homogenous porous media. In practice, budget, data, time, personnel skill sets, and regulatory expectations frequently drive the choice of modeling methods.

Recommendation 6. Develop appropriate hydrostructural conceptual models for fracture and rock matrix geometries and properties, and perform preliminary calculations (e.g., analytic or simple numerical) to better inform and allocate resources for site characterization, modeling, and remediation.

It is good practice to follow a systematic and well-documented approach to hydrostructural conceptual model development to understand the nature of underlying assumptions and simplifications, especially in the absence of supporting data. This can begin with generic geologic and geomechanical conceptual models and parameterizations that are informed but not limited by site-specific direct measurements. Next, develop a broad, semi-quantitative understanding of the important processes and parameters, and refine that understanding with simplified calculations that scope, define, and assess alternative models, their uncertainties, and the appropriateness of simplified versus more sophisticated modeling approaches. It is important to take advantage of such scoping calculations when making engineering decisions intended to reduce uncertainties in analysis—such as decisions related to site characterization activities and infrastructure or remediation design. Qualitative field data can be integrated into models through analysis tools such as parameter estimation, sensitivity analysis, and uncertainty quantification.

NUMERICAL MODELING

Commercial and publicly available numerical codes can facilitate fractured rock modeling for site-specific applications. To simplify calculations and make computations less resource intensive, these models often must be processed (upscaled) to average, narrow, or focus the ranges of specific hydrostructurally important properties and features.

Recommendation 7. Base numerical models on an appropriate hydrostructural model, ensuring that simplification and upscaling of the hydrostructural model maintain those features, properties, and processes that dominate contaminant transport, fate, and storage. Evaluate impacts of uncertainties introduced by simplification and upscaling.

Modelers need to consider carefully the appropriate applications and limitations of numerical models and analysis tools when selecting among them. The best models are those that synthesize data, confirm or refute the validity of the conceptual model, and quantify the range and uncertainties of expected system behaviors at the multiple scales appropriate to the problem of interest, using an approach consummate with the level of detail of the available data. It is important to apply equivalent continuum models only when the problem scale is much greater than the scale of the fractured rock mass. It is also important to analyze the need for risk- and uncertainty-based modeling approaches, even where calibrated and conditioned models match available measurements. Use non-deterministic modeling techniques (e.g., probabilistic and stochastic methods) when groundwater paths and response functions cannot otherwise be identified.

REMEDIATION AND MONITORING

The lack of a common framework, understanding, or expectations regarding remediation objectives, assessment, and realistic end points hinder effective engineering. Long-term remediation goals for most U.S. hazardous waste sites are based on drinking water Maximum Contaminant Levels (MCLs). Remediation can take decades or centuries to reach those MCLs. Regulatory agencies are beginning to accept long remediation time frames, and alternative approaches to remediation are becoming more acceptable.

Traditional pump-and-treat remediation methods draw water from fractures, remove contaminant, and return the water to the ground. Because contaminant is generally stored in the rock matrix of fractured rock, remediation focused solely on treating water in fractures is often futile. However, other remediation methods may be viable. Bioremediation, for example, relies on in situ biotransformation of contaminants by microorganisms. Thermal remediation promotes volatilization of contaminants, which are then removed via vapor extraction. However, better control of heating and extraction processes and more accurate prediction of results are needed to move this approach beyond the experimental stage.

Information regarding innovative underground remediation approaches is publicly available (e.g., through the U.S. Environmental Protection Agency). Less available is information specific to fractured rock site remediation. Lessons learned from remediation and monitoring in this setting are poorly documented, resulting in duplicated efforts and wasted resources in practice.

Recommendation 8. Develop and communicate realistic expectations related to remediation effectiveness through realistic goal setting and through explicit consideration of uncertainties in design, realistic use of natural attenuation, comprehensive monitoring programs, and dissemination of performance data to the technical community.

Best practices for remediation can be advanced through publicly accessible practitioner-driven and government-facilitated research-level documentation that details the remediation technologies applied across a variety of fractured rock settings. Expectations need to include the assumption that remediation of both rock fractures and matrix will likely be necessary. Estimates of plume longevity based on sound characterization need to be incorporated into remedial action plans that include natural attenuation as part of remediation. Regulatory frameworks that set realistic remedial objectives and formalize the transition from active remediation to long-term monitoring are needed, as are better information transfer mechanisms within and between agencies responsible for addressing fractured rock issues.

Recommendation 9. Incorporate long-term performance into monitoring system design.

Monitoring systems need to be designed for durability and to accommodate long-term performance and data needs. Accommodations include those for the expected operational and maintenance requirements for the duration of the infrastructure being monitored; for new technologies that might be developed during performance monitoring; and for potential variations in climate, water levels, temperatures, and other site conditions. An effective monitoring system includes meaningful sampling frequencies to monitor trends, and allows feedback that informs monitoring strategies in response to new trends or findings. Efficient monitoring system design is site specific and designed to require the minimal amounts of analytes to quantify performance effectiveness given localized discrete pathways, contaminant storage in the rock matrix, and geologic heterogeneity and anisotropy.

Monitoring systems need to be able to store, manage, and allow access to data in meaningful ways. Research on automated data collection, archiving, and retrieval and on the triggering of alarms when data values surpass tolerance levels is recommended. Research is also recommended on alternative approaches for remote monitoring of field-assessable parameters that trigger additional sampling and analysis. Best-practice protocols need to be established and communicated to future practitioners responsible for monitoring systems decades or centuries after monitoring systems are put in place. Advances in science and technologies need to be incorporated—as they develop—into engineering practice.

DECISION MAKING

Engineering processes often involve the selection and implementation of a design early in the project development stage before an adequate conceptual site model is formed. Legal requirements and fixed-cost approaches favor linear project development in which the site is characterized, the engineering design is selected, and the project is implemented, all without the benefit of any new information gathered during successive steps. By using observational methods and adaptive approaches, new information could result in altered thinking about the site hydrostructural model and realistic engineered outcomes.

Recommendation 10. Use observational methods and adaptive approaches to inform engineering decisions made for fractured rock sites.

Adaptive and observational approaches to characterize and manage fractured rock sites are more effective than prescriptive, linear approaches. Risk-based and performance-based criteria are useful, where appropriate, in prioritizing the components of contaminated site management. Model assumptions concerning transport pathways can be tested and refined with characterization and monitoring data, as can chemical and biological processes that affect remediation. Systems analysis approaches can be used to integrate the value of information and to support decisions made under uncertainty.

1
Introduction

Fractured rock contamination remains one of the greatest challenges to groundwater protection and cleanup (e.g., Steimle, 2002). Efforts to characterize relatively homogenous, unconsolidated soils (e.g., homogenous sand aquifers) and to remediate contaminated groundwater within them have become increasingly routine in recent decades. However, characterization of fractured rock and remediation of contaminated sites have not advanced to the same level. Because of the discrete nature of transport pathways in fractured rock, approaches applied commonly in soils can be ineffective in fractured rock. Contaminants can be transported great distances and at relatively high velocities along discrete channels, and therefore it becomes necessary to characterize both the rock matrix and properties of fractures that control or affect possible contaminant transport or remediation.

Rocks fracture when tensile, compressive, and shear stress conditions exceed their mechanical strengths. Fractures then can be modified as a result of tectonic and thermal stress and strain, as well as physical, chemical, and biological processes such as erosion, dissolution or precipitation of minerals, and degradation from roots. Fractures can be of varying length, aperture, and spacing. Water flow through fractures is commonly distributed unevenly, so the hydraulic significance of individual fractures within a fracture set can vary greatly. Often relatively few fractures present are hydraulically significant (discussion of this topic is found in Chapter 3, Box 3.1). Fractures generally occur in patterns dictated by geology, geomechanics, and geochemistry. In depth understanding of the various coupled processes that control fracture formation and contaminant transport will enable better characterization, modeling, monitoring, and remediation of the fractured subsurface. This is the focus of this report. This chapter provides an introduction to the committee's task and a brief qualitative discussion of the technical issues for readers not as familiar with the topic. More technical descriptions are provided in later chapters.

THE CONTEXT

Society engages in many activities that cause contaminants to enter into the subsurface. Contaminants may be released into the subsurface at ground level or through surface waters. Others migrate from infrastructure installed or wastes buried in the shallow or deep subsurface. Contamination of groundwater can occur in all types of rock and with many types of contaminants from many sources. Italicized text in this chapter presents one hypothetical scenario of fractured rock contamination.

> *Imagine an underground tank containing a hypothetical organic contaminant (HOC). The tank, buried near the surface of a mesa of fractured sandstone, has an undetected leak. The HOC is released slowly, but easily enters the fractures in the sandstone beneath the leak. Some of the fractures are interconnected, and the HOC moves rapidly downward through the fracture network, eventually reaching groundwater. A large amount of the HOC diffuses from those fractures into the low permeability–high porosity matrix of the sandstone.*

Because this particular contaminant is denser than water and only minimally water soluble, it migrates downward through the water-filled fracture network as a separate liquid phase. Some HOC droplets enter and are stored in the largest pores of the surrounding rock. Some HOC dissolves in and flows with the groundwater, moving away from the HOC source. Dissolved HOC diffuses from the fractures and enters the porous sandstone matrix. Some HOC is broken down by microbes in the groundwater or degraded by abiotic chemical reactions. Some HOC molecules adsorb on fracture surfaces, infilling minerals and organic materials. These processes influence the amount and location of HOC storage.

Fractured rock systems include intact blocks of rock surrounded by fractures and perhaps other features such as dikes, brecciated layers, or features associated with karst (e.g., Hoek and Bray, 1981). The rock blocks in the system may be crystalline with nominal porosity (i.e., many igneous rocks), or granular with varying amounts of cementation and porosity (i.e., sedimentary rocks). Insufficient characterization of the fractured rock environment, and of fate and transport processes within it, can lead to engineering strategies that fail to meet expectations. Contaminants, in many cases, diffuse quickly from fractures into the surrounding low permeability rock matrix. Diffusion back out of the matrix, however, may occur over decades or centuries, and the matrix effectively becomes a long-term contaminant reservoir. Remediated water in fractures can be then recontaminated as contaminants diffuse back into the fractures. Adequate characterization of the fractured rock environment goes beyond that for the homogenous porous environment and includes accounting for spatial heterogeneities and processes such as dissolution/precipitation, reduction-oxidation reactions, and biodegradation within the different geometries of hydraulic importance. Monitoring and characterization information is required to identify fracture and matrix porosities, and contaminant fate and transport in both fractures and rock matrix.

Uncertainties in characterization data and interpretation can result in inadequate site conceptual models and numerical models of fluid flow and transport. They can impact the effectiveness of monitoring programs and hamper lifecycle planning, mitigation, and remediation activities. These translate into less optimal management of contaminated sites or of sites where contamination is possible given the presence of certain types of infrastructure.

THE COMMITTEE'S TASK

The leak in the storage tank is discovered. Scientists and engineers try to understand where the HOC has migrated. They need more information about flow, adsorption, and reaction properties of the rock as well as the extent of in situ microbial activity that can process the HOC in the subsurface. Collecting this information will take time and involve considerable expense. The science team needs to design an informative program to assess the HOC plume.

To clean up the HOC, contaminated groundwater is pumped from the ground for surface treatment. Persistent low levels of the HOC are found to remain in the groundwater regardless of the pumping duration. Contaminated water is removed from the fracture system, but the HOC stored in the rock matrix diffuses back into fractures and contaminates the clean water migrating through the fractures from outside the HOC-impacted zone. The science team considers how to model this behavior to forecast potential contaminant migration scenarios. Uncertainties remain in their models even after applying state-of-the-practice characterization techniques, and the magnitude of those uncertainties cannot be determined.

Nearly a decade has passed since advances in the state of the art and state of the practice in fractured rock characterization and remediation have been examined comprehensively with respect to changes in regulatory regimes for implementing national policies. This report presents the findings of such an examination by a study committee of the National Academies of Sciences, Engineering, and Medicine (the National Academies). The study was funded by the National Aeronautics and Space Administration, the U.S. Nuclear Regulatory Commission (USNRC), the U.S. Department of Energy (DOE). The funding agencies have some level of regulatory or management responsibility for sites at which contamination in fractured bedrock is possible or present at depths of hundreds of feet, and extending several thousands of feet laterally. USNRC and DOE also have a role in identifying new sites or types of facilities considered for the long-term disposal of high-level radioactive waste. Some proposals involve disposal of wastes at great depths (i.e., up to 5 kilometers). This report is intended to inform the regulatory regimes or policy related to different types of infrastructure and their impacts on fractured rock systems and considers the long-term legacy of future waste repositories.

The National Academies convened a committee to undertake this study. The Statement of Task provided to that committee is provided in Box 1.1. Based on discussions with the study sponsors, the committee focused its attention on naturally fractured rock (although there are unconsolidated media in which fractures dominate flow regimes). Given the breadth of the committee's task, the committee made the strategic decision to focus on issues in the area beneath the vadose zone and up to approximately 5 kilometers in depth. Geotechnical, geologic, and hydrologic issues relevant throughout the lifecycle of engineered facilities are addressed. Research directions are suggested that could improve the state of the art, and applications are suggested that could enhance practice, including where better scientific understanding could inform regulations, policies, and implementation guidelines related to infrastructure development and operations. The scope of this study does not include techniques used to intentionally fracture low-permeability rocks for the purpose of extracting hydrocarbons (hydraulic fracturing); those tend to be purpose-specific. Neither does the report include discussion of the infrastructure from which contaminants originate, but rather focuses on their potential impacts to the subsurface geologic and hydrologic environments during the infrastructure lifecycles.

The committee includes researchers and practitioners with expertise in areas such as geohydrologic site characterization, hydrogeology, site-scale geotechnical and hydrologic modeling, contaminant fate and transport modeling, geotechnical and geohydrologic monitoring, environmental engineering, remediation practices, and risk assessment (see Appendix A for the committee member biographies). Committee members relied on their own expertise, the expertise of speakers and guests invited to their meetings, and information gathered during a workshop they organized. Appendix B provides the agendas for the open sessions of the committee meetings and the public workshop.

BOX 1.1 Statement of Task

Geological and geotechnical characterizing, modeling, and monitoring of the subsurface are integral to safe, economical, and responsible development, maintenance, operation, remediation, and decommissioning of infrastructure related to energy, water, waste, and transportation. Modeling and monitoring fluid travel paths and velocities through subsurface fractures and matrix are among the most significant engineering challenges associated with these tasks. An ad hoc committee of the National Academies of Science, Engineering and Medicine will conduct a study to address issues relevant to subsurface flow and contaminant transport in fractured media, including low permeability and low porosity media, as well as in deep (3 to 5 kilometers) fracture systems. Subsurface characterization, modeling, monitoring, and remediation (SCMMR) issues applicable throughout the lifecycle of engineered facilities that have the potential to release contaminants and pose risk to groundwater quality will be considered. As part of its information gathering, the committee will convene a workshop to examine the state of art and state of practice in

- Subsurface fracture and matrix characterization, especially relevant geotechnical, hydrological, and geochemical properties, and the development of conceptual models;
- Detection of fluid and contaminant pathways and travel times;
- Detection and modeling of factors that affect changes in geotechnical and hydrological properties over time (e.g., decades to millennia), including thermal, hydrological, chemical, and mechanical processes;
- Groundwater and contaminant transport modeling, monitoring, and remediation, and how these can aid decision making during facility design, operation, remediation, and decommissioning;
- Early indicators (e.g., change in fracture properties, moisture levels, background chemistry) of system failures resulting in unintentional release of fluids; and
- Potential mitigation measures to eliminate or reduce adverse impacts of system failures and related releases to the environment.

SCMMR knowledge generated from energy industry practice as well as carbon sequestration investigations will be considered. The committee will issue a final report that will include findings and conclusions with respect to (i) where research and development could improve the current state of art in SCMMR, (ii) where incorporation of scientific and technical advances could enhance the state of practice in SCMMR, and (iii) where enhanced science-based understanding could inform federal regulations, policies, and implementing guidance.

PREVIOUS NATIONAL ACADEMIES STUDIES

Multiple National Academies studies address issues related to fracture flow, contaminant transport, and subsurface remediation. The report *Rock Fractures and Fluid Flow: Contemporary Understanding and Applications* reviewed methods and strategies to characterize fracture flow in use from the mid-1970s to the early 1990s (NRC, 1996). A 2001 study on conceptual models of flow transport in the vadose zone built on the 1996 report and addressed development and confirmation of conceptual models that focus on how fractures affect recharge to local groundwater systems through all types of bedrock (NRC, 2001). A third report (2004) reviewed technologies for source remediation in various formations including fractured rock and proposed protocols to aid decision making for remediation strategies. Other NRC reports have stressed the need to establish guidelines to increase long-term direct monitoring of waste containment systems (2007), and a recent report explores alternatives to current groundwater remediation practices at complex sites (2013). Several other NRC reports address subsurface remediation from multiple perspectives (e.g., NRC, 1994, 1999, 2000, 2003), including from that of the effectiveness of remediation approaches. The number of reports reflects the challenges associated with subsurface remediation, aggravated in fractured rock, and indicates that technical gaps in remediation practice remain in spite of many recent advances. Regulatory guidelines and policies are in need of review, and this report is intended, in part, to describe the states of the art and practice to inform those reviews.

THE HETEROGENEITY OF FRACTURED ROCK

The heterogeneity of fractured rock can require more detailed characterization and analyses than what is required for less variable geologic materials. Rock fracture patterns form in response to stress fields in the Earth's crust (natural or anthropogenic) and rock type (e.g., sedimentary or crystalline) combined with various geomechanical, geochemical, and geobiological processes (discussed in Chapters 2 and 3). The causative factors of those processes vary over space and time, so heterogeneities and anisotropy in rock and fracture properties will also occur at various scales—from micrometer (grain size) to kilometer (regional) scales. Factors that cause fracturing also influence fracture geometries and their variations—characteristics such as fracture length, aperture, orientation, spacing, extent, intensity—as well as the geometric properties of the rock matrix blocks. Fractures can occur in multiple orientations at a given location as a result of evolving stress fields in the Earth's crust, as illustrated in Figures 1.1 and 1.2. Fracture geometries can be planar where they form preferentially along or quasi-orthogonal to bedding, as in the relatively undeformed sedimentary rocks pictured in Figure 1.3, but they may be non-planar, as might occur in the folded sedimentary rocks pictured in Figure 1.4.

Fracture geometry is a controlling factor of flow: fractures create discrete pathways for both groundwater flow and contaminant transport. Fractures can be highly conductive, but they can also be flow barriers, for example, where faults contain clay gouge (e.g., Cain et al., 1996; Crawford et al., 2008). Exploration boreholes drilled at a site may not capture features that control contaminant fate and transport, as is discussed in more detail in Chapter 5.

FIGURE 1.1 Fractures at many scales and orientations can occur at a single site. The satellite image shows the superposition of multiple large-scale fracture sets (lineaments) of different orientations in the quartz monzonite of Joshua Tree, California. The scale line in the lower right-hand corner represents 1 km (T. Engelder 2013 workshop presentation). SOURCE: Imagery © 2015 Google, Map data © 2015 Google.

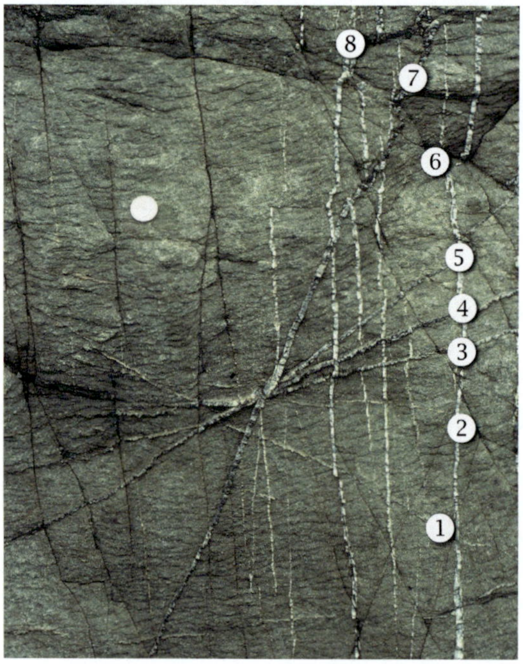

FIGURE 1.2 Filled fractures (veins) in an Ordovician clastic rock showing the complexity of stress field in the crust in New South Wales, Australia. The orientation of the veins (numbered 1 through 8) and fractures indicates rotation of the stress field of approximately 120° over time (coin in upper left-hand corner for scale). The nearly vertical veins and fractures cross cutting the other fractures represent a more recent stress field. SOURCE: T. Engelder 2013 workshop presentation.

FIGURE 1.3 Surface outcrop of mudstone host rock for trichloroethylene contamination at the Naval Air Warfare Center in New Jersey. These fractures are controlled by bedding planes and jointing that is nearly perpendicular to bedding. SOURCE: U.S. Geological Survey. Available at http://toxics.usgs.gov/photo_gallery/photos/nawc/NAWC_outcrop1.JPG.

FIGURE 1.4 The once horizontal bedding planes of this mudstone from Fox Island, Alaska, have been deformed and folded. Some fractures coincident with folded bedding planes are traced in red. Some fractures roughly orthogonal to bedding planes are shown in green. SOURCE: U.S. Geological Survey. Available at http://alaska.usgs.gov/science/geology/images/photo_gallery.html.

CONTAMINANTS IN FRACTURED ROCK

A variety of contaminants can be detrimental to groundwater quality and the environment. Given their different characteristics, distinct strategies are needed to prevent, mitigate, monitor, and remediate contamination of fractured rock. Indeed, proper understanding of specific contaminants and their interaction with fractured rock is critical to reduce the uncertainties associated with their management. Specific details regarding individual contaminants are beyond the scope of this report, but below are brief discussions of common water soluble and non-aqueous phase liquids. Vapor phase contaminants are also found in fractured rock settings, but are more common in the vadose zone.

Water Soluble Contaminants

Contaminant solubility in groundwater can range from fully soluble to less than parts-per-million levels of solubility. Fully miscible contaminants (e.g., nitrates, acetone, and methanol) readily dissolve in water and can travel with the groundwater great distances from their sources. Dissolved contaminants may react with surrounding geologic materials, precipitate from solution, or exchange electrons or ions with the rock surface and change oxidation states. These reactive dissolved contaminants migrate more slowly than the transporting water and are said to be "retarded" or "attenuated." Some attenuated contaminants—particularly organic chemicals such as benzene—tend to be sorbed and held by organic matter found in fractures or by the rock matrix (Zytner, 1994). Many weakly soluble organic contaminants (e.g., benzene) can be biodegraded under the right circumstances by naturally occurring microorganisms in fractured rock environment (e.g., Johnston et al., 1994).

Some dissolved contaminants are radioactive and inherently unstable. Like other soluble contaminants, their rate of movement is a function of their chemical properties. Some radioactive dissolved contaminants such as tritium have a short enough half-life that over a period of a few decades, radioactive decay leads to significant reductions in concentration. Others, such as components of spent nuclear fuels, have half-lives spanning many thousands of years.

Non-Aqueous Phase Liquids

Liquids that do not dissolve in water in appreciable amounts are referred to as non-aqueous phase liquids (NAPLs). NAPLs such as dry-cleaning liquids, petroleum products, and cleaning solvents contaminate sites throughout the world. Many NAPLs have components that dissolve in water (e.g., benzene and toluene from gasoline at low concentrations). NAPLs migrate through fractured rock in liquid phases separate from water and flow at different rates (and potentially directions) than water, depending on their viscosities, relative densities, and surface tension with water.

Light non-aqueous phase liquids (LNAPLs; e.g., gasoline) are less dense than water and accumulate primarily above the water table, but can be transported horizontally and vertically in fractures beneath the water table (Adamski et al., 2005). LNAPLs can spread or be trapped beneath and above the water table with changing groundwater levels. This can cause contamination of remediated groundwater when it comes in contact with mobile or residual LNAPL (Newell, 1995). Dense non-aqueous phase liquids (DNAPLs; e.g., trichloroethene) are denser than water and sink below the water table to locations that can be difficult to identify. Most chlorinated solvents are DNAPLs. These include effective but sometimes toxic degreasing agents used widely for industrial cleaning, dry cleaning of clothes, and other applications. DNAPL groundwater contamination can be most difficult to identify and manage and can often go undetected at a site given the variables that affect its transport and storage (Huling and Weaver, 1991). Because DNAPL is denser than the groundwater, it sinks until it encounters a barrier. Figure 1.5 demonstrates the potential distribution of DNAPLs in a karst setting.

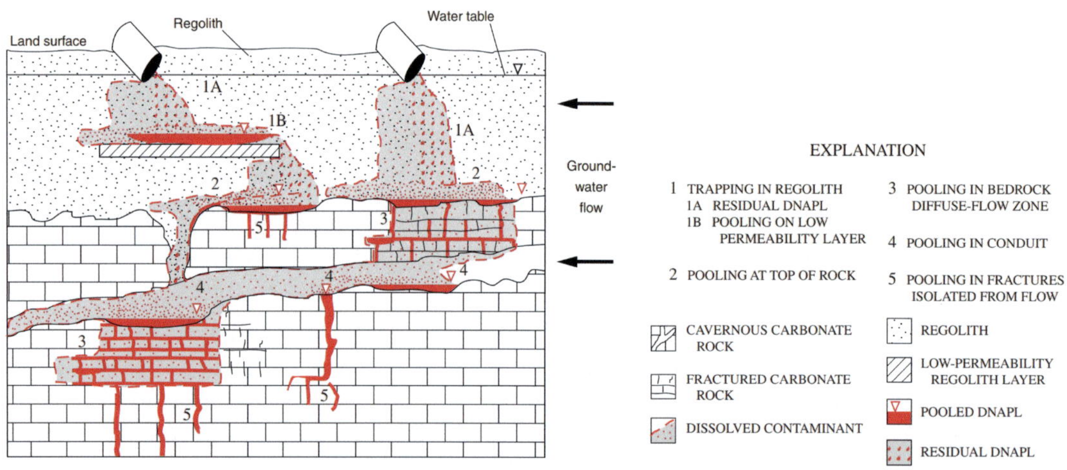

FIGURE 1.5 Potential distribution of DNAPLs in a karst setting. In this case, DNAPL is released into the vadose zone and migrates downward to fractured bedrock until a barrier is reached (not shown). SOURCE: CLU-IN (Contaminated Site Clean-Up Information). Available at http://clu-in.org/contaminantfocus/default.focus/sec/Dense_Nonaqueous_Phase_Liquids_(DNAPLs)/cat/Overview.

FATE AND TRANSPORT

Depending on local geologic, geomechanical, and hydrogeologic conditions, individual fractures or fracture networks can become high-velocity discrete pathways for groundwater and contaminants. Advective travel times in these pathways may be significantly shorter than flow through lower permeability rock matrix materials (Fossum and Horne, 1982). Different contaminant concentrations, velocities, and geochemical and biologic conditions may be found within fractures and matrix.

Advection, addressed throughout this report, is the transport of dissolved chemicals, particles, or dissolved or volatilized gas by a carrier fluid (i.e., groundwater) as it flows through the subsurface. The transport rate and flow direction of substances is not necessarily the same as the velocity and direction of the carrier fluid. As a contaminant from a source (i.e., a storage container) is carried downstream (a plume), the contaminant spreads laterally and vertically (hydrodynamic dispersion) as a result of rock fracture and matrix geometries and local variations in flow velocity. The measure of the spread of contaminant as it travels through the groundwater is the hydrodynamic dispersion. Other factors that affect the behavior of contaminants include fracture network geometry, variations in fracture aperture and roughness, and the local hydrodynamics where fractures intersect (Neuman, 1990). This is discussed more fully in Chapter 3.

Diffusive transport occurs in response to random molecular motion. At the macroscale, diffusive transport reflects concentration gradients and the contaminant moves from zones of higher to lower concentration. Chemical constituents of fluids in fractures initially diffuse quickly into fluids in the rock matrix surrounding fractures because concentration gradients are high. They diffuse more slowly back into the fractures from the matrix because the gradients are much smaller and are present in two directions (away from and toward the fracture). This slow release of chemicals from the rock matrix may occur over long periods (decades to centuries for some substances).

The fate and transport of chemicals in fractured rock are also affected (sometimes profoundly) by chemical and biological processes. Chemical alterations can include precipitation of dissolved constituents from solution, dissolution of solid constituents into groundwater, sorption of chemicals onto the surfaces of rock and into the rock matrix, changes in oxidation state that affect mobility or toxicity, volatilization from liquid to gaseous state, dissolution from gaseous state into liquid state, and chemical transformations into different compounds. These chemical changes, in turn, can be affected by changes in dissolved oxygen, pH, temperature, and other parameters. Biological processes also affect these chemical processes in fractured rock sites as microorganisms, and their byproducts can act as catalysts for chemical reactions. This can result in contaminant transformations, precipitation/dissolution reactions, gas bubble nucleations, and changes in pH and redox conditions. Microorganisms can also form flocs (aggregates of particles) and biofilms that modify fluid flow and may result in clogging. Biological processes are especially critical in terms of biotransformation of organic and inorganic contaminants in the subsurface. In fractured rock formations, biological processes may be significant, depending on the specific site, at both shallow depths and at depths of many kilometers.

KNOWLEDGE FROM RELATED AREAS

This report focuses on the understanding of fractured rock in the context of understanding the behavior of contaminants and contaminant remediation. However, there has been significant research on fluid movement in rock for applications in the energy sector, including that related to oil and gas exploration and production, geothermal energy production, and carbon capture and storage (CCS) (Zimmermann et al., 2011; Rutqvist, 2012; Kim and Moridis, 2015). These energy-related activities occur at depths from 1 to 4 kilometers, so the knowledge gained from research in those areas informs the main topics of this report. Whereas this report does not include lengthy discussion of research and development in the energy sector, relevant references from that sector are provided throughout this report.

REPORT ORGANIZATION

This report examines recent progress and addresses issues raised since the publication of the comprehensive 1996 NRC report *Rock Fractures and Fluid Flow: Contemporary Understanding and Applications*. Progress has been made characterizing, modeling, monitoring, and remediating the fractured rock environment, but significant challenges remain. These are due, in part, because only a small number of fractures often carry most of the flow in a rock mass. Without understanding the geologic context of a site, it can be difficult to identify and locate those hydrologically important fractures and understand how contaminants may be transported and move between rock fractures and matrix. Modeling the interactions between concurrent physical, chemical, thermal, and biological processes that determine contaminant migration is inherently complex.

Subsequent chapters of this report examine progress in these and other areas related to the Statement of Task:

- Chapter 2 provides background on the characteristics of fractured rock important to contaminant migration.
- Chapter 3 considers fundamental phenomena underlying coupled processes that occur in fractured rock and control the fate and transport of contaminants within fractures, and between the fractures and rock matrix.
- Chapter 4 explores the importance of appropriate hydrostructural conceptual models and the representation of physical processes in contaminant migration models in fractured rock systems, including inherent limitations. The goal of a modeling effort is to examine how, where, and how fast contaminants might move in the subsurface; examine the impact of potential efforts to remediate contaminated sites; and inform where and which characterization and remediation efforts may be most efficient and cost-effective.
- Chapter 5 reviews methods for acquiring information about the subsurface in ways that support conceptual and quantitative modeling of the hydrostructural system and contaminant transport and storage and that determine the effectiveness of engineered systems.
- Chapter 6 explores options available for site remediation and management and provides a set of important considerations for remediation of fractured rock.
- Chapter 7 considers decision-making processes for characterizing, monitoring, and remediating fractured rock sites.

Findings and conclusions are found throughout this report; however, Chapter 8 synthesizes the committee's overarching findings and recommendations. These stress the importance of interdisciplinary approaches to understanding fractured rock and the interactions and processes within it that control contaminant transport and fate; the importance of developing appropriate hydrostructural conceptual models and applying numerical models that accommodate them; the need for developing appropriate goals for fractured rock site remediation; and the importance of adaptive and observational approaches. Recommendations in this report are written to improve science, engineering, and research. An integrated systems approach for engineering fractured rock sites is emphasized, so recommendations are presented in a broader systems context.

2

Physical Characteristics of Fractured Rock Controlling Flow and Transport

In fractured rock, individual rock fractures, blocks of rock matrix, and fracture infilling can each have discrete porosities that need to be understood to identify those features that control flow, storage, and transport. Characteristics of those features need to be well enough understood to discern what of the rock's void space can be considered, from a hydrological engineering point of view, to be active or inactive. Matrix and fracture porosities contribute differently to contaminant fate and transport, and although the importance of multiple porosities has long been recognized within academe, much engineering and hydrogeologic work in general practice is dependent on tools that cannot accommodate multiple porosities. This chapter introduces the geologic, geomechanical, geochemical, geobiological, and hydrogeologic contexts to understand and effectively characterize rock fracture and matrix geometries and properties. This information provides the necessary basis for understanding fate and transport processes (described in Chapter 3), modeling (described in Chapter 4), characterization (described in Chapter 5), and remediation of fractured rock (described in Chapter 6).

This chapter begins with a discussion of qualitative aspects of rock fractures. The chapter then focuses on the quantitative description of rock fracture and matrix porosities, including for example, the concept of "sets" of similar fractures, descriptions of fracture orientation, extent, intensity (number of fractures in a given area), spatial pattern, and infillings. The final section focuses on how understanding fracture genesis and evolution over geologic time can assist understanding fate and transport in fractured rocks. The geologic, geomechanical, and geochemical processes that cause fractures to form and evolve allows the geometry and relevant characteristics of fractures and rock blocks to be predicted and quantified.

QUALITATIVE FRACTURE DESCRIPTION

From a rock mechanics viewpoint, a fractured rock mass is a system that includes intact blocks of rocks surrounded by fractures and other discrete features such as dikes, brecciated layers, and features related to karst (e.g., Hoek and Bray, 1981). Fractured rock systems, however, can vary greatly geologically, geomechanically, geochemically, geobiologically, and hydrogeologically. Thinking of fractured rock as a system is appropriate when considering fluid flow, storage, and transport in the rock. Characteristics of one part of the system can affect fluid behavior in other parts of the system, so it becomes necessary to identify and understand the discrete features that control behaviors of interest. Fracture flow and transport properties are sensitive to in situ stress through shear and normal displacements, producing changes in fracture aperture and roughness. They are also sensitive to geochemical conditions through sorption and precipitation processes. Such processes are discussed further in Chapter 3.

An individual fracture can be characterized as a separation of the rock ("face"). It consists of two or more surfaces with variable spacing. The space is described as a physical aperture and is generally filled with some combination of gases, liquids, and geologic materials. The geologic materials can consist of gouge, breccia, coatings, and minerals placed or formed in the fractures by chemical, advective, biological, or geomechanical processes. Fracture aperture is further discussed in Box 2.1. The surfaces themselves are rarely simple planes and need to be characterized in terms of their roughness, planarity, and undulation. Local roughness features are referred to in rock mechanics as asperities. Fracture persistence refers to continuity of the fracture—the percentage of the fracture surface that has a non-zero aperture (i.e., it does not include infilled fractures) (Ulusay and Hudson, 2007).[1] Individual fractures can be defined by the detailed three-dimensional coordinates of all of the faces. However, it can be more convenient to describe individual fractures by their location (center), orientation, size (extent in different directions), and physical aperture distribution.

BOX 2.1 Fracture Aperture

Fracture aperture can vary by orders of magnitude, from tightly closed to several centimeters in breadth, and can be measured or estimated by different means that yield different results. Aperture values are most useful when reported as arithmetic or geometric means with corresponding variability or confidence bounds. Field data show that fracture aperture typically follows a log-normal distribution, with the standard deviation $\sigma[\log(h)]$ often exceeding 0.4 (Stone, 1984; Dverstorp and Andersson, 1989; Dverstorp et al., 1992; Nordqvist et al., 1992). Although it is common practice to refer to a single "fracture aperture," different physical processes control the relationships between the fracture aperture as it might be seen in outcrop or a core and flow capacity (e.g., fracture hydraulic conductivity), transient flow behaviors (e.g., fracture stroativeity and diffusivity), and solute transport (e.g., residence time, dispersivity). It is therefore useful to define distinct aperture measures for each of these with defined functional relationships between these measures (Bear et al., 2012).

- Geometric (mechanical) aperture is determined by the visible opening of the fracture in the field, core, or image logs (Evans et al., 1992; Ishibashi et al., 2012; Bisdom et al., 2015);
- Hydraulic aperture is determined by back calculation of steady state flux measurements (Chen et al., 2000; Baghbanan and Jing, 2007);
- Storage aperture is determined by back calculation of transient flow measurements (Rutqvist et al., 1998; Nielsen, 2007); and
- Water residence time aperture is determined by back calculation of solute transport measurements (Bear et al., 2012).

In particular, the hydraulic aperture inferred from measured flux depends on the geometry of the fracture surfaces and may not agree with the geometric aperture (Schwartz and Smith, 1988; Iwano and Einstein, 1995). Therefore, the popular descriptor "fracture aperture" is most informative if accompanied by its spatial variability, fracture roughness, dissolution, and infilling (Reimus and Callahan, 2007). Variations in fracture aperture arise from geomechanical process (e.g., fracture propagation processes, shear displacement, in situ stress history) and geochemical processes (e.g., precipitation, dissolution, and weathering). The resulting fracture aperture pattern can range from tensile fractures with limited contact area to fractures that have been almost completely infilled through mineralization and remain open only at fracture intersections.

---

[1] The term "persistence" has conflicting usages in the rock mechanics literature, including uses to indicate fracture size, intensity, and continuity. This report refers to persistence in terms of fracture continuity, in a manner consistent with Ulusay and Hudson (2007).

Fracture channels form where portions of fractures are blocked preferentially (e.g., by mineralization or local asperity contacts) and where portions of fractures are locally more permeable (i.e., due to localized dissolution, mechanical processes at fracture intersections, or mismatches between asperity patterns on fracture faces). Fracture channels, as discussed below, can be important because they influence groundwater residence times and reactive surface areas. Multiple fractures within a rock mass can be described most effectively by grouping fractures with similar spatial patterns, extents, and orientations into sets. Physical characteristics such as roughness, infillings, physical apertures, and planarity can then be described for the fracture sets as variability distributions among the fracture population and correlations between properties (e.g., between orientation and extent, or between extent and physical aperture).[2]

QUANTITATIVE FRACTURE DESCRIPTION

The ability of fractures in rock aquifers to transmit groundwater will vary over many orders of magnitude (Freeze and Cherry, 1979), as confirmed in transmissivity measurements[3] of individual or closely spaced fractures in boreholes in various rock types (Shapiro and Hsieh, 1998; Novakowski et al., 2000). Changes in transmissivity in fractured rock aquifers can also be abrupt, so hydraulic properties of fractures cannot be assumed to be uniform. For this reason, it is important to develop an appropriate geomechanical framework of fracturing for use as the basis to infer groundwater flow pathways. Transmissivity is further discussed in Chapter 3.

Fracture set spatial patterns can be quantified by defining the functional variation in fracture intensity and the interactions between fractures (e.g., Rogers et al., 2015). The intensity of a given set may be lower if near another set, or if the fracture orientation of a given set may be defined at a specific orientation relative to that of an adjacent fracture set. Fractures terminating against other fractures can control the fracture size. The mean spacing between fractures can be used to parameterize fracture intensity; however, for many practical and theoretical reasons, fracture intensity is better formulated as volumetric intensity (Dershowitz et al., 2000). The spatial pattern of fractures can also be parameterized in terms of termination percentages between fractures of different sets (Ulusay and Hudson, 2007) and by fracture intersections (Dershowitz, 1985).

Figure 2.1 illustrates a continuum of fractured rock intensity. At one extreme (see Figure 2.1a), the fractured rock can have very high fracture intensity. When characterizing such intensity in a site conceptual model (see Chapter 4), this type of system sometimes can be assigned a single porosity, as would a homogenous media or a soil. At the other extreme (see Figure 2.1c), there may be very few fractures of hydrogeologic significance, and those fractures would need to be included explicitly and deterministically in a site conceptualization, just as they would be in a conventional faulted porous media model. Both the fractures and matrix need to be adequately characterized in a site conceptualization for fractured rock environments between these two extremes (see Figure 2.1b). For such systems, there are numerous discrete fracture pathways to consider explicitly or stochastically, and there can be large differences between fracture and matrix flow velocities. Such interactions between fracture and matrix porosities affect transport. Box 2.2 lists fractured rock characteristics other than aperture (see Box 2.1) that affect contaminant fate and transport.

---

[2] The extreme tails of the distributions can be important for flow and transport and may necessitate large sampling sizes to characterize properly.

[3] Transmissivity is the volume of flow per unit time per unit width.

FIGURE 2.1 Photos demonstrating the continuum of fracture intensity. Highly fractured rocks such as the fault breccia shown in (a) might behave much like unconsolidated media (note the lens cap circled in red for scale), while rocks with low fracture intensity (c) may have few fractures of hydraulic significance (spacing between faults and fractures may be 50 feet). Rocks with intermediate fracture intensity such as those shown in (b) are more complex to conceptualize given the contrasts (spacing between fractures may be uneven and up to several inches). SOURCES: (a) Park Administration of the National Park Northern Velebit; (b) Robin Stott; and (c) OMCV.

BOX 2.2 Characteristics Affecting Fate and Transport

The following is a list of some characteristics that affect fate and transport (fracture aperture, also important, is described in Box 2.1). These characteristics are influenced by mechanical, geochemical, and geologic processes (see Chapter 3). Mineralization from in situ fluids, for example, may reduce porosity (Emmanuel and Berkowitz, 2005). Geomechanical processes that result in changes in rock porosity or fracture aperture are also important.

Fracture roughness. Roughness and aperture variability affect flow conditions within the fracture. Surface irregularities and roughness can be analyzed using fractals (Brown and Scholz, 1985, 1986; Power and Tullis, 1991; Brown, 1995; Odling and Roden, 1997). Smaller-scale roughness can be measured using fracture profilometry (e.g., Keller and Bonner, 1985; Durham and Bonner, 1993; Ameli et al., 2013). However, the small-scale aperture variations measured often are not related to the hydraulic, transport, and storage aperture patterns that control flow and transport (Hjerne et al., 2010).

Fracture intensity and extent. Fractures exist over multiple dimensions, from pervasive microfracturing to large discontinuities in the rock extending over hundreds of meters. The physical abundance of fracture void space does not necessarily dictate the viability of using effective properties to characterize groundwater flow and contaminant transport over a given physical dimension (Snow, 1970; Herda et al., 1991). Fractures can cross other fractures and exhibit significant size, or stop at an existing fracture. Intersecting fracture angles can range from acute to normal.

Fracture infillings. Fractures can be filled with granular or cementitious gouge formed by localized mechanical erosion processes (i.e., primary fines with the same mineralogy as the matrix rock), or by chemical dissolution of the parent rock and reprecipitation in the form of new minerals (i.e., secondary fines, typically clays such as kaolinite, illite, and montmorillonite) (e.g., Mahmoudzadeh et al., 2013).

Rock matrix porosity and pore size. The intrinsic matrix porosity of the rock itself (including void space and dead-end fractures) may not contribute to fluid transport significantly, but it has a critical role in storage, coupled diffusion-advection, and bioactivity. There is increased awareness of the role of rock matrix porosity, especially in problems of contaminant transport (Zuber and Motyka, 1994).

Rock matrix pore size distribution. Porosity alone is not a complete descriptor of the void space in the rock matrix. Pore size distribution needs to be adequately characterized to understand immiscible fluid invasion and the potential for bioactivity (Mitchell and Santamarina, 2005).

Description of Rock Matrix

Similar to fractures, rock matrix between fractures and other discontinuities are characterized by a combination of geometric (e.g., block size, shape, and spatial pattern), hydraulic, and transport properties (e.g., hydraulic conductivity, porosity, storativity, tortuosity, minerology, and sorption characteristics). Matrix properties also vary spatially due to geologic, geomechanical, geochemical, and geobiological histories and current in situ conditions (Bibby, 1981; Berryman and Wang, 1995). Fracture properties also evolve (Tsang, 1984; Long and Witherspoon, 1985). Boxes 2.1 and 2.2 include rock matrix characteristics that may contribute to fate and transport in fractured rock. Igneous and metamorphic rocks have minimal matrix porosity that can range from less than 1 to 2 percent of the rock volume (e.g., Dietrich et al., 2005). Sedimentary rocks typically have higher matrix porosities, ranging as high as 30 to 45 percent of the rock volume. Table 2.1 includes one set of values derived for total and effective intact rock porosity and hydraulic conductivity. Other values are also used (e.g., those developed by the Argonne National Laboratory).[4]

Rock mass storage capacity can be underestimated if hydraulically inactive fracture porosity is not considered. From a transport standpoint, void space in rock can be divided into three types of porosity: mobile, immobile, and inaccessible. Mobile porosity is the percentage of the total volume of void space (matrix and fracture) that contributes to advective migration—this can be considered active void space. Immobile porosity is the combined volume of interconnected matrix porosity and hydraulically inactive fracture porosity divided by the total volume. Inaccessible porosity is the void volume of the rock not accessible by fluids or diffusing compounds, divided by the total volume. Hydraulically inactive fractures can become reservoirs for immobile fluids and contaminants. The increased storage can dominate contaminant transport in some rock types, particularly where the matrix porosity is low. The importance of immobile porosity in contaminant transport in fractured rock systems cannot be overstated. Continued diffusion from mobile to immobile porosities results in changes in the distribution of contaminant contained within a rock system. Contaminants may be contained initially in the mobile porosity, then migrate to be found primarily in the immobile porosity. Errors in travel time predictions, concentrations in hydraulically active fractures, and plume sizes can result from poor understanding of immobile porosity.

IMPORTANCE OF FRACTURE GENESIS

Fracture geometry and properties are determined by the combined geomechanical, geochemical, and geobiological processes that control fracture genesis and fracture evolution. It is therefore critical to understand the geologic setting to predict the location of fractures and their properties. NRC (1996) provides a good review of the relation between underlying geology and fracturing. Ways in which geology, geomechanics, and geochemistry influence fracture patterns relevant to fate and transport are described in Box 2.3.

TABLE 2.1 Intact Rock Porosity (Total and Effective) and Hydraulic Conductivity

Rock	Total Porosity n (%)	Effective Porosity n_e (%)	Hydraulic Conductivity K (m s^{-1})
Granite	0.1	0.0005	3×10^{-14}–2.0×10^{-10}[a]
Limestone	5–15	0.1–5	1.0×10^{-9}–6.0×10^{-6}
Sandstone	5–15	0.5–10	3.0×10^{-10}–6.0×10^{-6}
Shale	1–10	0.5–5	1.0×10^{-9}–6.0×10^{-6}

[a] Hydraulic conductivity value for "unfractured igneous and metamorphic rocks;" total porosity and effective porosity are for crystalline granite.
SOURCE: Domenico and Schwartz, 1990.

[4] See http://web.ead.anl.gov/resrad/datacoll/porosity.htm (accessed August 21, 2015).

BOX 2.3 Influences of Geology, Geomechanics, and Geochemistry on Fracture Geometry

Orientation

Rock fractures are created primarily when tensile (i.e., leading to expansion) or shear stresses, or their combination, exceed the strength of the rock. Tensile fractures generally form perpendicular to the minimum stress, while shear fractures generally form at an angle of approximately 20–35 degrees relative to the maximum stress direction (Dershowitz and Einstein, 1990). Knowledge of stress orientations can lead to understanding the origin of at least some of the major fracture orientations.

Spatial Pattern and Intensity

Lithology exerts strong control in bedded volcanic, metamorphic, and sedimentary rocks. Many bedded rocks (e.g., sedimentary rocks such as sandstone) include fracture systems perpendicular to bedding, with the major set of fractures perpendicular to the minimum horizontal stress at the time of deposition. In addition, extensive bedding plane parting fractures can form at lithologic contacts (Narr and Suppe, 1991).

The spatial pattern and intensity of fractures within folds can be defined by the unique patterns of the fold crest, forelimb, and backlimb (e.g., Stearns, 1967). Fracture geometry and properties can be derived and fracture measurements can be understood within the context of the structural position (i.e., where the fractures are relative to the fold crest, forelimb, and backlimb). The local fracture orientations can also be related in terms of the local coordinate system of bedding, as bed-perpendicular fractures remain bed-perpendicular when stratigraphic layers are folded and faulted.

Flow Geometry (Channeling)

Flow and transport geometry within fractures is a combination of the effect of hydraulic boundary conditions and the geometry of the connected pore-space within each fracture and at fracture intersections (Winberg, 2010). The connected pore-space geometry arises from the combination of the mating of asperities on fracture faces to form a roughness profile and the occurrence of fracture infillings such as breccia, gouge, and precipitated minerals (e.g., Casini et al., 2012). Flow and transport channelization is therefore dependent on the geologic, geomechanical, and geochemical history of the fractured rock. Examples of these processes include the increased fracture surface roughness symptomatic of fractures formed by tensile mechanisms, as well as dissolution and fracture infilling that cause constricted pathways within only a small portion of fracture area. In carbonate rocks, dissolution processes associated with flowing groundwater frequently form open conduits and caverns that dominate the hydrogeology (MacQuarrie and Mayer, 2005; Singurindy and Berkowitz, 2005; Ellis et al., 2013).

Size

Lithologic controls play a lesser role in intrusive igneous rock masses. For these rocks, local and regional stress and strain distributions define the fracture spatial pattern, including orientation, size, shape, and intensity. Important constraints on fracture size and intensity can be related to termination of fractures at fracture and fault interactions. Variation in intensity can frequently be related to distance from the hanging wall and footwall damage zones of faults, and local stress redistribution due to faulting.

Shallow Fracture Patterns in Soils

Fracture geometry in fractured soils can be dependent on many of the same geochemical, geologic, and geomechanical processes that control fracture geometry in rocks, or they may be controlled by processes unique to soils (e.g., desiccation cracking). The same principles used to predict fracture orientation, intensity, size, spatial pattern, and hydraulic properties in fractured rock may apply to fractures found in soils including residual soils (e.g., Arnould, 2006). Desiccation cracking is well known in clay-rich soils with high water content, including compact clay-rich materials (Kleppe and Olson, 1985). Saprolite residual soils can be expected to have fracture spacing and orientation consistent with that of the parent unweathered rock. Similarly, where near surface weathering diminishes with depth, fractures observed in residual soils can be expected to change as residual soils transition to unweathered fractured rocks at depth (Christensen and Mooney, 1995; Rudnick and Fountain, 1995).

3

Flow and Transport: Underlying Processes

Chapter 2 reviewed the physical and structural properties of fractured rock that are relevant to subsurface contaminant fate and transport. This chapter reviews processes that affect fate and transport over different relevant timescales, including hydrologic, thermal, chemical, biological, and mechanical processes. These properties and processes form the basis for the conceptual and computational models used in the analysis of fractured rock sites discussed in Chapter 4.

ADVECTION AND DISPERSION

Advection, dispersion, and diffusion in fractured rock are the key processes that control the fate and transport of contaminants in fracture networks and contaminant migration into and out of the rock matrix. The underlying physics of these processes are the same in both soils and fractured media; however, their relative impacts and the degree of control of the solid structure (granular versus cemented) are significantly different when considering fractured rock settings. Some important additional considerations unique to fractured rock are discussed in the following sections. For the most part, the following processes are invariant with depth below ground surface.

Impacts of Transmissivity and Flow Focusing on Advection

As discussed in Chapter 2, groundwater flow in individual fractures is controlled by the geometry and connectivity of those fractures. Advective transport relies on the bulk motion of a fluid (i.e., groundwater) driven by the total energy gradient. Contaminants (which can include dissolved species, suspended solids, bacteria or viruses, and colloids) are transported by advection in flowing groundwater. Fractures are generally the primary paths for advective flow (Birkholzer et al., 1993), although the rock matrix may play an important role in more permeable rocks such as some sandstones.

Transmissivity is defined as the volume of fluid per unit time (flow) per unit width for a hydraulic gradient equal to 1.0. It is a common descriptor of the hydraulic property of fractures and varies by many orders of magnitude among fractures. The analysis and discussion in Box 3.1 demonstrates that relatively few of the fracture population (i.e., less than 10 percent) transports more than 90 percent of the total advective flow in most situations.

The identification of flow-controlling fractures is crucial to the development of a reasonable conceptual model for fractured rock sites; hence, their identification is also a critical step in the decision-making process (see Chapter 7). In many fractured rock settings, advection is the predominant transport process, and thus a rigorous, quantitative discernment of highly transmissive pathways is required. In most fractured rock settings, averaged or effective media properties (the approach applied in sediments) do not represent the discrete nature of advection appropriately, and discrete fracture models are needed (see Chapter 4). Project objectives and fracture size characteristics relative to the physical dimensions of the phenomenon or site of interest may require, however, that effective media properties be used to represent fluid flow and contaminant transport in fractured rock (see Chapter 6).

BOX 3.1 Transmissivity Control on Advective Flow in Fractures

The distribution of transmissivity of fracture sets is typically lognormal (e.g., Fransson, 2002). The coefficient of variation (COV) of the distribution is defined as the ratio between the standard deviation of log(transmissivity) divided by the mean of log(transmissivity). The figure below shows the relationship between the number of fractures responsible for 90 percent of the total flux and the transmissivity COV. For very low variability in transmissivity (COV = 0.37), 10 percent of the fractures contribute 90 percent of the flow. For higher variability (COV = 0.72), only 0.01 percent of the fractures contribute 90 percent of the flow. Hence, flow concentrates or focusses along the most conductive pathways.

The figure below is based on an assumption of parallel fractures, where each fracture is defined by a single, homogeneous, isotropic transmissivity value, and the fracture network is fully connected. However, in natural fractured rocks, fracture connectivity is limited, transmissivity varies within a fracture plane (Tsang and Neretnieks, 1998; Neuville et al., 2011), and some flow spreads into neighboring fractures. Hence, while marked flow focusing remains in natural systems, it is less pronounced than anticipated by this analysis.

The analytical results demonstrated in the figure agree with numerical simulations of fracture networks (e.g., Frampton and Cvetkovic, 2011) and are supported by field observations (e.g., Shapiro et al., 2007; Follin et al., 2014). Furthermore, flow also channelizes within fractures in response to aperture variability associated with fracture roughness and mineralization, as well as network-scale flow focusing (Crawford, 2010). There are scale effects on conductivity and dispersivity (e.g., Neuman, 1990; Schulze-Makuch et al., 1998, 1999), but the control of flow by relatively few fractures remains.

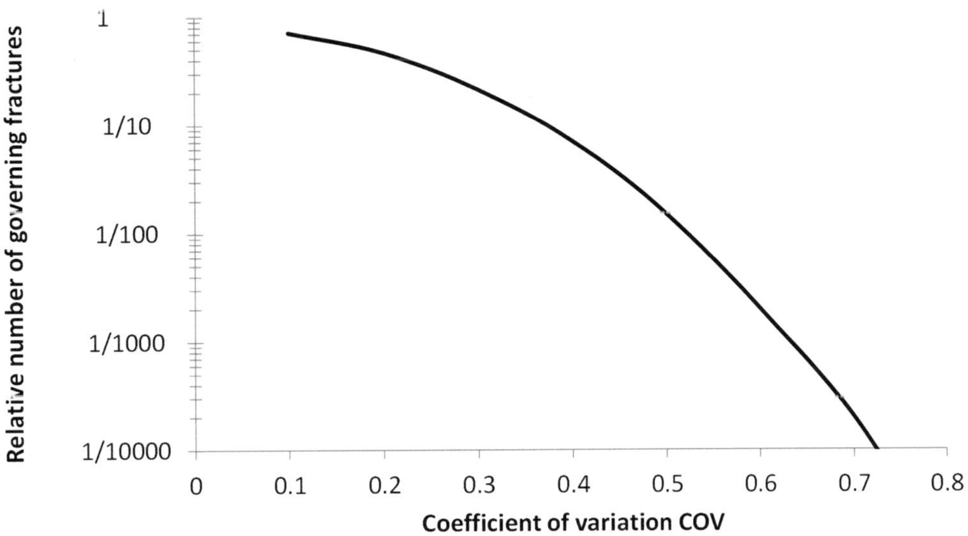

FIGURE Relative number of fractures responsible for 90 percent of the flow as a function of the transmissivity coefficient of variation COV defined as the ratio between the standard deviation and the mean of the log(T).

Structural Controls on Hydrodynamic Dispersion

Dispersion due to mechanical effects in flow pathways results from mixing at fracture intersections, fracture tortuosity (the length of the actual flow path relative to the linear distance between two points), aperture variability in a single fracture, transmissivity variation between fractures and ensuing velocity changes, and fracture interconnectivity. The effects of mixing/dilution in fractured rock due to structural controls (e.g., intersections, aperture variability) are much stronger than those due to similar controls in soils. Although dispersion occurs primarily in two dimensions (within the fracture plane), the effects are often observed in three dimensions because of interconnectivity. Representation of dispersion using classical dispersion tensor approaches used in soils is not appropriate, but the concept of a three-dimensional assemblage of two-dimensional features should be considered. Hydrodynamic dispersion impacts residence times in fractured rock and is likely scale dependent (Berkowitz and Scher, 1995; Rubin and Buddemeier, 1996; Dagan and Neuman, 1997; Berkowitz, 2002; Neuman, 2005). Figure 3.1 shows that hydrodynamic dispersion is most pronounced near sources and plateaus quickly once the scale of observation approaches that of the scale of the fractures. This behavior contrasts the classical assumption of a continuous increase in dispersion with increasing travel length used in granular media. Recent work challenges the classic paradigm of advection in soils (Payne et al., 2008; Hadley and Newell, 2014) and is more in line with the mechanistic description of fractured rock.

The prevalence of flow focusing and structural controls on dispersion in fractured rocks (as compared to sediments) is an inherent and critical characteristic in fractured rock masses and has a pronounced effect on all flow-related problems, and equivalent continuum soil-derived approaches need to be used with caution and appropriate qualifications.

Dispersion and Transport in Cyclic Flow

Another transport mechanism results from the combination of hydrodynamic dispersion and cyclic fluid flow whereby inflow-outflow cycles (with zero remnant invasion at the end of the cycle) combine with pore-scale mixing to render effective transport (see Figure 3.2): even though the cumulative fluid flow volume is null at the end of each inflow-outflow cycle, solute is effectively transported into the medium (Knothe Tate et al., 1998; Wang et al., 2000; Goldstein and Santamarina, 2004; Claria et al., 2012). The more dominant structural controls in fractured rock make this phenomenon potentially more important than in soils. Efficient transport in cyclic fluid flow can affect natural systems (such as nearshore saltwater intrusion due to tidal action) and can be used to engineer treatment strategies where monotonic advective regimes cannot be established but where inflow-outflow pumping cycles can be imposed.

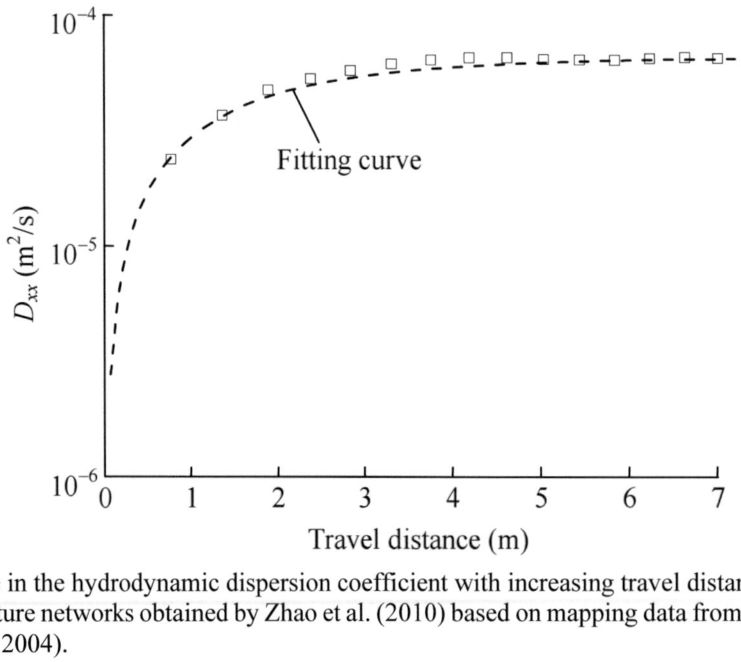

FIGURE 3.1 Change in the hydrodynamic dispersion coefficient with increasing travel distance. Hydrodynamic dispersion results in fracture networks obtained by Zhao et al. (2010) based on mapping data from Sellafield in the United Kingdom (Min et al., 2004).

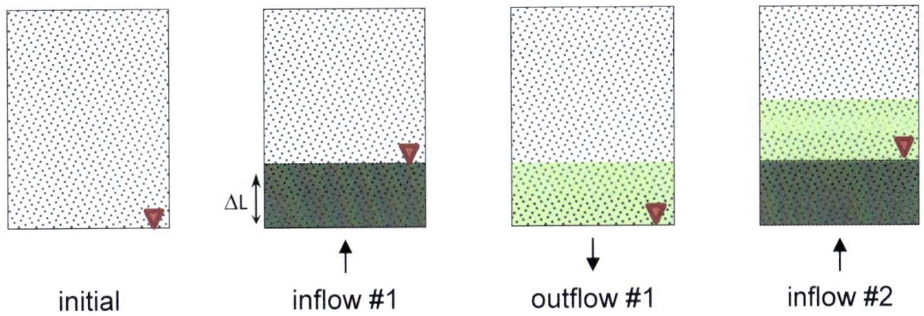

FIGURE 3.2 Accelerated diffusion-like response during inflow-outflow cycles due to mixing and pore-scale hydrodynamic dispersion. Fluid with contaminant invades the water-saturated rock from the bottom and reaches a height, ΔL, shown with the inverted red triangle (inflow #1; dark green); then it recedes, leaving some contaminant behind (outflow #1; remnant contaminant shown as light green). During inflow #2, the high concentration contaminant reaches height ΔL once again (dark green); however, the remnant contaminant from the previous invasion is displaced upward (light green advances to higher position).

HYDRO-MECHANICAL COUPLING

The fluid pressure field present in geologic systems affects the effective stress field, and changes in effective stress affect fracture apertures, which in turn modify the fluid pressure field. These processes are depth invariant. The greater the effective compressive stress, the smaller the aperture. For example, if a well is pumped, the water level in the well declines, the fluid pressure in the rock is reduced, the effective compressive stress increases, apertures close, and transmissivities decrease (Segura and Carol, 2008). The significance of this trend depends greatly on the compressibility of the fractured rock.

The reverse situation is of great importance: if a well is pressurized, hydro-mechanical coupling can lead to the localized opening of preexisting discontinuities in the fractured rock (Zoback et al., 1977; Warpinski and Teufel, 1987; Beugelsdijk et al., 2000; Chuprakov et al., 2011; Fisher and Warpinski, 2011). The ensuing distortion band dilation of the surrounding medium can explain the high-flow efficiencies observed in prestructured media, such as the recovery of gas from shales following hydraulic fracture enhancement (Dusseault, 2011). Hydro-mechanical coupling gains relevance under high pumping regimes, such as in engineered hydraulic fracturing.

CHEMICAL PROCESSES: DIFFUSION AND REACTION

Although fractures dominate advective transport in most fractured rock, the volume of void space in the rock matrix is, in most cases, orders of magnitude greater than the volume of void space in the fractures. This skewed volume ratio results in the matrix accounting for the majority of the contaminant storage in fractured rock settings. Under these circumstances, diffusive contaminant transport between the matrix and the fractures exerts considerable control on the spatial distribution of the contaminants, as well as timescales of transport.

Diffusive Matrix-Fracture Coupling

Chemical constituents dissolved in groundwater can diffuse between fluid-filled rock matrix and fractures as a result of random molecular motion, the direction of concentration gradients, and the tortuosity of the pore space. Diffusion is almost unaffected by pore size (first order approximation); hence, diffusion takes place within and between the matrix and fracture porosities (e.g., Neretnieks, 1980; Maloszewski and Zuber, 1993). The relative rates of advective and diffusive transport are dependent on fracture and matrix properties (see Box 3.2). In most cases, advection is the main transport mechanisms along fractures, while diffusive transport prevails in the matrix and is orthogonal to the fracture plane. The relative timescales for matrix diffusion and advective transports along fractures have a critical role in the development of site conceptual models, in the design of site investigations and in the selection of remediation strategies for fractured rock systems.

Fracture-Matrix Diffusive Interaction During Contaminant Advection

The diffusion of contaminant into the matrix during advection along fractures results in contaminant mass retardation as contaminant is "lost" from the advection-dominated fractures into the matrix (Golubev and Garibyants, 1971; Tang et al., 1981; Sudicky and Frind, 1982; VanderKwaak and Sudicky, 1996). Figure 3.3 demonstrates that actual contaminant velocities can be significantly lower than the advective velocity within the fracture.

The end result of the rapid diffusion of contaminant into the rock matrix is that a significant amount of contaminant will reside within the matrix in a relatively short timeframe. The relative percentage of chemical contaminants in the matrix versus total contaminant for different fractured systems (where contaminant enters entirely as dissolved compounds in fractures) is shown in Figure 3.4 (VanderKwaak and Sudicky, 1996). In this example, most of the chemical mass is in the matrix within 1 year of it entering a fractured rock system. Understanding the importance of matrix diffusion and its timescales are critical for planning site investigations in fractured rock systems.

Matrix diffusion takes place at all depths. Given the geothermal gradient, analyses need to incorporate increasing diffusion rates and decreasing fluid viscosities with increasing temperatures. Both diffusive and advective processes are also affected by the tendency of deep aquifer rocks to have lower matrix porosities and narrower fracture apertures.

Delayed Release: Back Diffusion

The rock matrix porosity acts as a reservoir that can sustain long-term release of contaminants back into fractures through reverse or back diffusion (Parker et al., 1994; Reynolds and Kueper, 2002; Lipson et al., 2005). This back diffusion reflects the same diffusive process that occurred during the original contaminant invasion of the matrix. High concentration gradients during advection lead to a fast increase in contaminant concentration and storage in the matrix (and the non-conductive fractures within the rock matrix). Gradually lower concentration gradients (which may occur during remediation activities) result in remediation times that can exceed hundreds or thousands of years (Reynolds and Kueper, 2002; Lipson et al., 2005; see Figure 3.5).

BOX 3.2 Definition of Peclet's Number and Timescales for Diffusive and Advective Transport

The relative rates of advective and diffusive transport are captured in Peclet's number:

$$\mathrm{Pe} = \frac{\text{Advective transport rate}}{\text{Diffusive transport rate}} = \frac{vl}{\mathrm{D}}$$

The diffusion timescale, $t_{dif} \approx B^2/(4D)$, varies with the square of the characteristic length (i.e., half the block size B [m]), and it is inversely proportional to the diffusion coefficient D [m^2/s]. The timescale for advection, $t_{adv} = B/v$, depends on the travelled distance B [m] and the flow velocity v [m/s]. Advection along fractures and matrix diffusion adds complexity to the fate of contaminants in fractured rock masses. More information about diffusion timescales can be found in Parker et al. (1994).

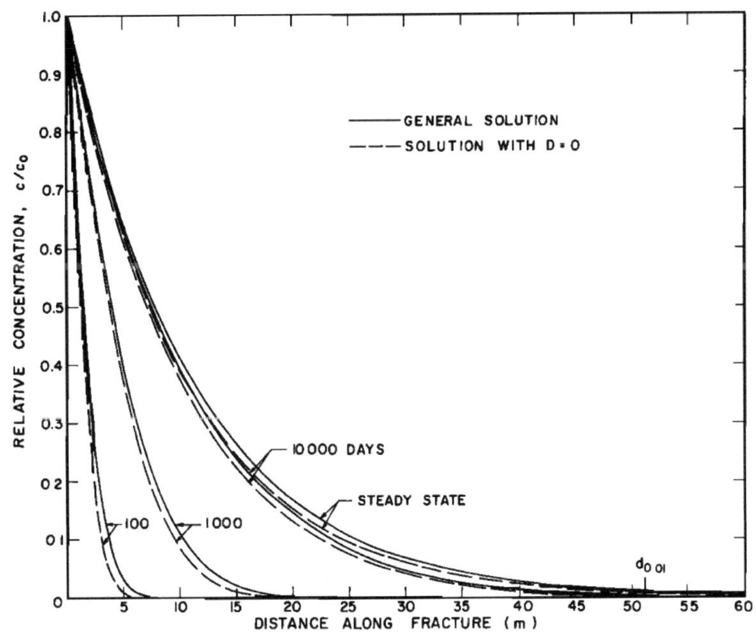

FIGURE 3.3 Contaminant transport retardation during advection as species diffuse into the rock matrix. Note that the fluid velocity in the fracture is 0.75 m/day. A constant concentration of contaminant enters the left side of a fracture (distance = 0 m) in an infinite rock block (perpendicular to the plane of the fracture) at a constant velocity (0.75 m/day). The center of mass of the contaminant in the fracture (C/Co) is 0.5 after 1,000 days of flow is at a distance of 5 meters from the inlet, resulting in an apparent retardation factor (v*t/d) of 750 m/5 m = 150. SOURCE: Tang et al., 1981.

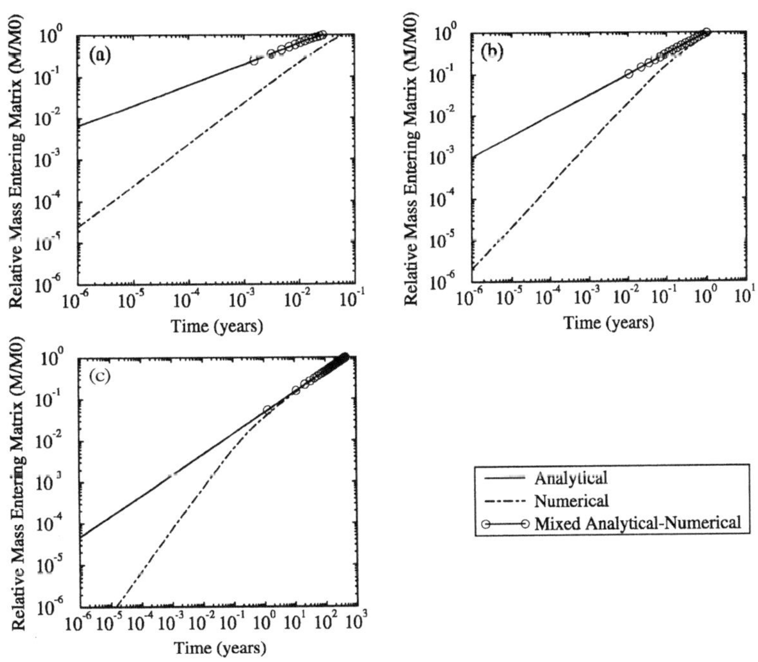

FIGURE 3.4 Relative mass in matrix versus time (initial mass = Mo = 0 in the matrix) for three different fractured media: (a) clay, (b) sandstone, and (c) granite. For matrix porosity, ϕ is the primary parameter varied between the three panels in the figure (ϕ clay = 0.35 > ϕ sandstone = 0.1 > ϕ granite = 0.006). Lines represent different numerical approaches to the calculation. SOURCE: VanderKwaak and Sudicky, 1996.

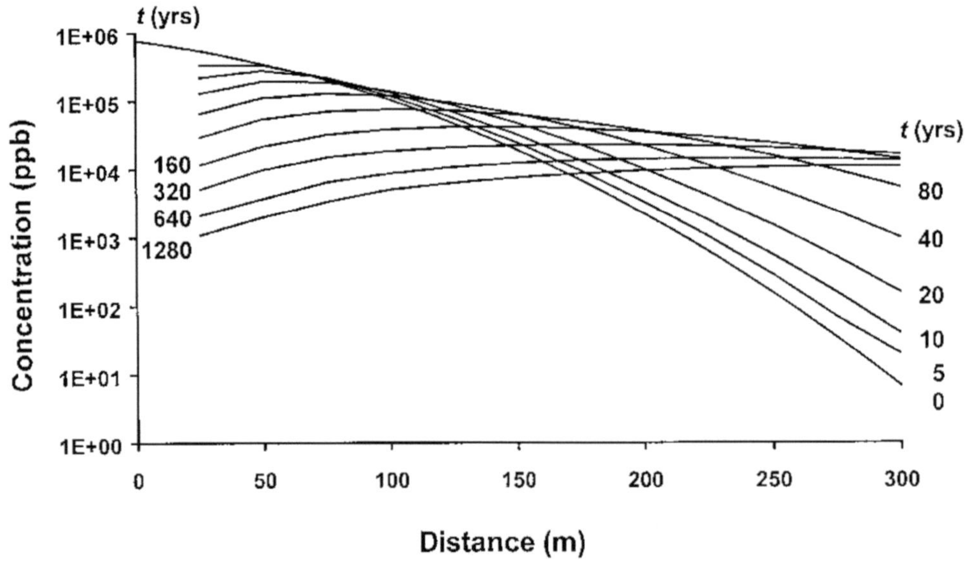

FIGURE 3.5 Concentration in a fracture downgradient of a contaminant source (located at distance = 0 meters) demonstrating back diffusion from storage following complete removal of contaminant source. Immediately following removal of the source (t = 0) concentrations are highest near the source (distance = 0). Concentrations near the source decline with time but remain above 1,000 ppb for more than 1,000 years. SOURCE: Lipson et al., 2005.

Time-dependent matrix back diffusion effects on plumes are not well characterized in complex fracture networks; yet, plume attenuation in fractured rock systems is critically important for decisions related to risks that plumes pose to human and ecological health (Sudicky and Frind, 1982; McKay et al., 1993; Lipson et al., 2005). The complexity of fracture networks greatly hinders the development of quantitative relationships that could estimate back diffusion contaminant mass loading to systems.

Reactive Fluid Transport

Geochemical reactions within fractures can exacerbate flow heterogeneity and may lead to the formation of pipes, channels, and caverns (Tsang and Tsang, 1989; Dagan and Cvetkovic, 1996; Cvetkovic et al., 1999). As described in Chapter 2, some dissolved contaminants may precipitate out of solution, plug flow paths, and force transport through different pore spaces. The opposite can also occur when dissolution of minerals by flowing groundwater opens up previously plugged pore spaces. In addition, biological processes can alter fluid transport by causing precipitation of chemicals or biomass growth and cause a reduction in the area available for fluid flow. Recent publications provide experimental evidence (Fredd and Fogler, 1998, 1999; Singurindy and Berkowitz, 2005; Yasuhara et al., 2006; Noiriel et al., 2007; Elkhoury et al., 2013; Ruiz-Agudo et al., 2014) and complementary numerical studies (Chen and Doolen, 1998; Chaudhuri et al., 2008; Szymczak and Ladd, 2009; Chaudhuri et al., 2012; Ameli et al., 2014) relevant to fractures and fracture networks.

Geochemical reactions, advective contaminant transport, and diffusive transport rates control the evolving topology of dissolution (Steefel and Lichtner, 1994; Szymczak and Ladd, 2013). Discerning the relationships between the reaction rates and advective and diffusive transport rates guides the selection of the scale at which reactions must be incorporated into a site conceptual model. The probability of localized dissolution increases in settings with high reaction rates in advection-dominant transport (see Box 3.3). The formation of a dissolution feature alters the flow regime, and other dissolution features will not form nearby (i.e., an exclusion distance between dissolution channels emerges). Localized dissolution channels extend the zone of high fluid pressure from the fracture inlet into the rock mass and lower the effective stress in the fractured rock mass. Eventually, large-scale instabilities may follow.

BOX 3.3 Relationship Between Chemical Reaction Rates and Advective and Diffusive Transport Rates

The relationship between chemical reactions and advective and diffusive contaminant transport rates can be captured in two dimensionless numbers: Peclet's number, defined in Box 3.2, and the Damköhler number

$$\mathrm{Da} = \frac{\alpha l}{v} = \frac{\text{Reaction rate}}{\text{Advective transport rate}}$$

where v is flow velocity (in meters/second), l is the characteristic length scale (in meters), and α $[1/v]$ is the mass exchange coefficient. Experimental and numerical studies show that the dissolution topology can be anticipated from Pe and Da numbers (Figure). In particular, dissolution localization develops at high Da>10^{-2} and Pe>10^{-2} numbers.

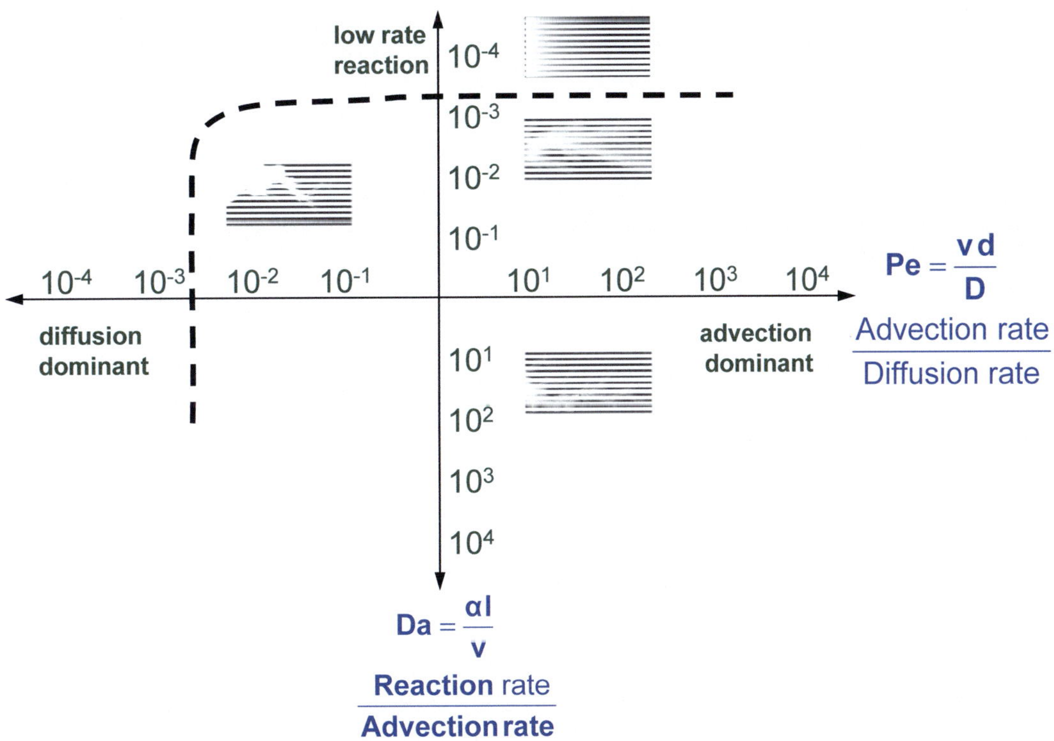

FIGURE Regimes in reactive fluid transport. Advective and diffusive transport couple with reaction rate to cause various dissolution morphologies. The lighter colors in the photographs inserted in this plot show the zones where mass dissolved. The dotted lines suggest the boundary for localized dissolution based on numerical and experimental results. SOURCES: Fredd and Fogler; 1998; images of experimental results from Golfier et al., 2002.

Chemical re-precipitation within the rock mass often complements dissolution (Dobson et al., 2003). Precipitated species may be the same as those dissolved upstream (e.g., frequent examples in carbonates), or selective precipitation-leakage results in other mineral formation (e.g., clay minerals). Fracture-scale dissolution and re-precipitation reactions can have pronounced macro-scale effects. The potential for reactive fluid transport, therefore, should be considered in conceptual model development.

The long-term, large-scale projects envisioned for carbon capture and storage emphasize the need to understand and model carbon dioxide (CO_2) storage capacity and the behavior of CO_2 as it migrates through the reservoir along various pathways. Critical issues regarding fluid flow in reservoirs deeper than 1 kilometer have been identified through work at long-term CO_2 storage sites such as the Sleipner field offshore Norway. Eiken et al. (2011) suggest the need to understand and model combined diffusion-convective mechanisms and fluid movement and to understand the mechanisms (such as diffusion, dilution, buoyancy forces, and capillary trapping) occurring when supercritical CO_2 is injected in the reservoir.

BIOLOGICAL PROCESSES

Microorganisms inhabit almost every environmental niche in the biosphere, including shallow and deep fractured rock environments. Within these environmental niches, they form complex communities that can act as catalysts for a wide range of chemical reactions, including nucleation of gas bubbles that change water saturation; precipitation of minerals that cement fractures; dissolution of minerals that widen fractures; degradation of toxic organic contaminants; and formation of biofilms that hinder fluid flow and lead to clogging (Baveye and Valocchi, 1989; Kindred and Celia, 1989; Taylor and Jaffe, 1990a,b; Taylor et al., 1990; Clement et al., 1996; Ehrlich, 1999; Ross et al., 2001; Wagner et al., 2013).

Biological Rock Mass Properties: Conditions for Life

Life requires sufficient space, water, nutrients, energy, and adequate environmental conditions to flourish in rock masses. The presence and health of microbes in the subsurface generally is constrained by pore diameter (e.g., Fredrickson et al., 1997; Rebata-Landa and Santamarina, 2006; Bartlett et al., 2010). The nominal size of microorganisms is about 1 micrometer (see histogram of identified bacterial and archaeal cells versus cell size in Figure 3.6). Consequently, microbial activity should be anticipated in most fractures, even at depths in excess of hundreds of meters. However, the presence of life in the pores within the rock matrix may be limited by the absence of pores greater than or equal to 1 micrometer (Phadnis and Santamarina, 2011). Rocks such as shales, carbonates, evaporites, and granites often have mean pore sizes less than the nominal 1 micrometer cutoff (see Table 3.1).

The activities of microbial communities in fractured rocks can exert significant effects on both physical and geochemical characteristics and can have a critical role in natural attenuation. The potential role and importance of microorganisms on natural attenuation in fractured rock environments remain poorly recognized, even at a theoretical level within the research community. Furthermore, although some techniques are used currently to quantify rates of natural attenuation via biological processes in porous media (e.g., Illman and Alvarez, 2009), there exists almost no practical guideline to estimate those rates at fractured rock sites.

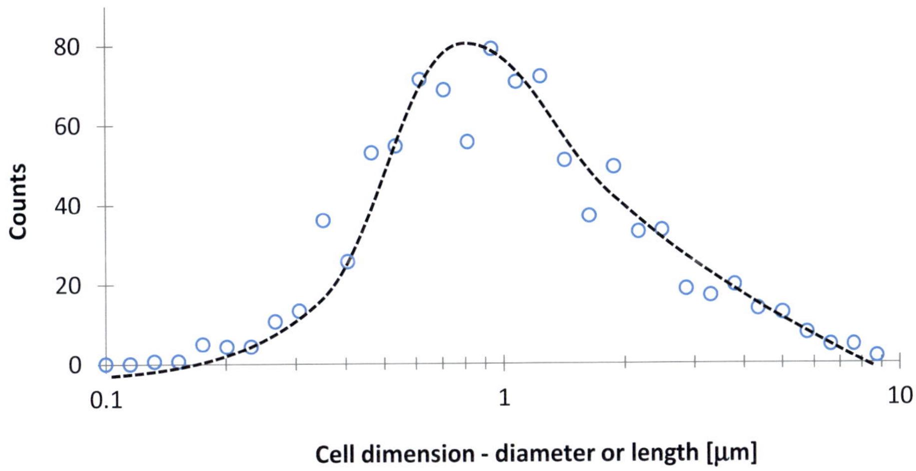

FIGURE 3.6 Histogram of identified bacterial and archaeal cells versus cell size. Cells have a minimum size in order to contain all components and organelles required for life. While diameters can be as small as ~0.2 µm, a nominal 1 µm bacterium size can be selected for first order analyses. Data compiled from more than 100 publications on bioactivity in sediments.

TABLE 3.1 Hydraulic Conductivity and Pore Diameters for Different Rock Types

		Pore Diameters (µm)			
Rock	Hydraulic Conductivity (m/s)[a]	min.	med.	max.	Reference
Dolostone	$10^{-5.5}$ to 10^{-9}	–	10.17	–	Tanguay and Friedman, 2001
Limestone	$10^{-5.5}$ to 10^{-9}	–	31.5	–	Laine et al., 2008
Sandstone	$10^{-5.5}$ to 10^{-10}	5	–	100	Verges et al., 2011
Shale	10^{-9} to 10^{-13}	0.005	0.1	0.75	Loucks et al., 2009
Unfractured Igneous Rocks	10^{-9} to 10^{-14}	< 0.0038	0.413	80	Mosquera et al., 2000[b]; Dultz et al., 2013

[a] Hydraulic conductivity data from Freeze and Cherry (1979).
[b] Unfractured igneous rock pore diameter data are from granite rock samples.

Although biodegradation within the primary porosity of the rock matrix has been hypothesized as a natural attenuation mechanism that may reduce back diffusion loading to fractures, little is known about the presence, health, or activity of microbial communities in rock matrices.

Common Bio-Mediated Geochemical Cycles

Microbial communities can participate in the biogeochemical cycling of many elements, including carbon, nitrogen, sulfur, and numerous metals (Taylor and Jaffe, 1990a; Clement et al., 1996; Krumholz et al., 1997; Ehrlich, 1999; Geller et al., 2000; Ross et al., 2001; Wagner et al., 2013). For example,

- Solid phase ferric iron oxy-hydroxides can be mobilized to soluble ferrous iron form by iron reducing bacteria. This reaction generally requires the presence of an electron donor such as hydrogen or organic carbon.
- Microbial communities also can reduce soluble sulfate to sulfide that readily forms precipitates with cations; this important biogeochemical process affects flow through fractured media.

- Microbial archaea (methanogens) reduce CO_2 to methane and bacterial acetogens reduce CO_2 to organic acetate using hydrogen as an electron donor. High methane concentrations can result in the formation of bubbles that may also affect fluid flow.

Consequently, when groundwater transports compounds such as iron, manganese, sulfur, and CO_2 from suboxic to oxic environments or vice versa, chemolithotrophic organisms[1] can flourish, oxidizing or reducing dissolved forms to generate less soluble forms that can become clogging precipitates, dissolving insoluble forms to widen fractures, and generating gases that become bubbles. The geochemical environment within both fractures and matrix affects the potential for biological reactions. Analyses need to consider spatial and temporal variations in geochemistry, particularly at shallow depths where recharge and anthropogenic impacts are probable.

MIXED-FLUID CONDITIONS

Mixed-phase fluid conditions are common in fractured rock; examples range from vapor-water conditions in the vadose zone and deep geothermal reservoirs, to water-organics phases in contaminated sites and oil reservoirs, to vapor-liquid conditions that emerge during thermal and biomediated remediation strategies. In particular, many common organic contaminants exhibit very low miscibility in water (e.g., non-aqueous phase liquids [NAPLs]) and exist as a separate phase to water, as described in Chapter 1. Other contaminants that are lighter than water, such as light NAPLs (LNAPLs), released at or near land surface impact the water quality in fractured rock, particularly where there is little or no soil cover between the ground surface and the fractured rock. Because LNAPLs are lighter than groundwater, they infiltrate the vadose zone, driven by the fluid pressure in the LNAPL body and the effects of gravity. Furthermore, LNAPLs can be found up to several meters below the water table in systems subjected to water table fluctuations, or when a shallower LNAPL pool above the water table drives the localized LNAPL invasion into fractures and macropores (Adamski et al., 2005). On the other hand, when injected into water or brine-saturated formations at depth, LNAPLs will float and pool against seal layers, laterally restrained by geometric traps or by capillarity—the same behavior that occurs in the injection of CO_2 for geologic storage.

Dense NAPLs (DNAPLs) such as trichloroethylene (TCE) have higher density and often lower viscosity than water. DNAPLs tend to migrate downward through the vadose zone, reach the water table, and displace water in the fractures (Kueper and McWhorter, 1991; Kueper et al., 1993). DNAPLs have the potential to migrate to great depths in fractured rock environments, and vertical migration is a critical part of understanding a fractured rock site impacted by DNAPLs.

Under these mixed fluid (or multi-phase) conditions, the interfacial tension at the contact between the two fluids will sustain a pressure difference between them (capillary pressure) that is inversely proportional to the curvature of the interface. When present in fractures, the curvature of the interface must be such that the interface can exist within the fracture; therefore, fracture aperture controls capillary pressure and the distribution and invasion potential of an immiscible phase.

NAPLs behave similarly to air in the subsurface: an air bubble can only invade a water-saturated porous material if the interface between water and air is forced into the porous material. Similarly, NAPL can only penetrate a fracture if the capillary pressure is great enough to deform the interface to the scale of the aperture. Once within the fracture, migration will occur along the

[1] Chemolithotrophic organisms are those that use reduced inorganic compounds for energy.

path of least resistance (typically, the largest aperture pathway) and can stall if aperture sizes decrease to the point where the available capillary pressure cannot deform the interface to the required degree. Once within a fracture, NAPL will only penetrate the surrounding matrix pores if the pore size is large enough, as will be the case in many sandstones. Box 3.4 demonstrates the interplay between the physics of capillarity (determined by interacting fluids and the mineral substrate) and the geometric characteristics of fractures. It should be recognized, however, that NAPLs can dissolve in water to levels that are orders of magnitude above regulatory criteria. For example, TCE, one of the most ubiquitous organic contaminants found in the United States, has a solubility in water of approximately 1,000 mg/L (e.g., Russell et al., 1992) and a regulatory limit of approximately 0.005 mg/L. Thus, even if TCE is not found as a separate phase in the rock matrix, it may be present in the rock matrix at concentrations that are a risk to human and ecological health.

The relationships presented in Box 3.4 lead to the following general observations related to mixed fluid conditions in fractured rock masses:

- Capillary forces
  - Affect contaminant invasion; variations in fracture aperture and matrix pore size result in a complex contaminant distribution in fractured rocks.
  - Have a secondary role in the mechanical behavior of fractured rock masses unless the pore size, d, is sufficiently small so that capillary suction approaches the overburden stress σ; from Laplace's equation (Equation a in Box 3.4), $d \leq 2, T_s/\sigma$.
  - Affect the displacement of fines in fractures subjected to mixed-fluid conditions, as fines are subjected to gravity (Archimedes buoyant weight), electrostatic forces (van der Waals and double layer repulsion as in DLVO theory), Stokes drag, and capillary forces (Sarkar and Sharma, 1990; Santamarina, 2002; Weisbrod et al., 2002).
- A non-wetting fluid (i.e., NAPL) will advance along fractures and will not likely penetrate into small pores in the rock matrix as a separate phase. Thus, NAPL storage will take place primarily in fracture porosity, although, as mentioned previously, the NAPL may have dissolved and is present in water contained in the rock matrix at concentrations not exceeding the solubility limit in water.
- Migration is not necessarily regular as a stable front; indeed, common field conditions lead to viscous fingering (high contrast in fluid viscosity) or capillary fingering, as shown in Figure B in Box 3.4.
- Under constant capillary pressure, an invading non-wetting fluid (i.e., fluid A in Box 3.4) will migrate along the wider interconnected regions of a fracture; hence, the rock matrix will retain a high saturation of the wetting fluid (i.e., fluid B in Box 3.4) even after the non-wetting fluid percolates across the entire rock mass (i.e., breakthrough).
- After invasion, buoyancy and capillarity define fluid distribution under equilibrium conditions.
- In smaller pores or narrower aperture fractures, viscous forces caused by advective flow (natural or pumping-induced) may not overcome capillary forces and contaminants will remain "capillary trapped" (this is termed "residual" NAPL).

BOX 3.4 The Interplay Between Physics of Capillarity and the Geometry of Fractures

Interactions between capillarity, viscous drag, and buoyancy in mixed fluid flow govern migration and distribution patterns. Consider fluid A invading a fracture filled with fluid B, whereby fluids A and B exhibit limited mutual solubility (e.g., gas and water; oil and water). Intermolecular interactions between fluids A and B, and between the fluids and the rock, result in interfacial tension T_s [N/m] and the emergence of a characteristic contact angle θ between the fluid interface and the rock surface (see Figure A). The pressure difference ΔP required for a non-wetting fluid A to invade a fracture of aperture d_{frac} filled with the wetting fluid B can be estimated from Laplace's equation for a fracture plane:

$$\Delta P = \frac{2T_s \cos(\theta)}{d_{frac}}$$

in terms of the surface tension T_s between the fluids and the contact angle θ formed by the fluid interface and the mineral. This equation highlights the preferential migration of the non-wetting fluid into wider fractures rather than the smaller pores in the matrix.

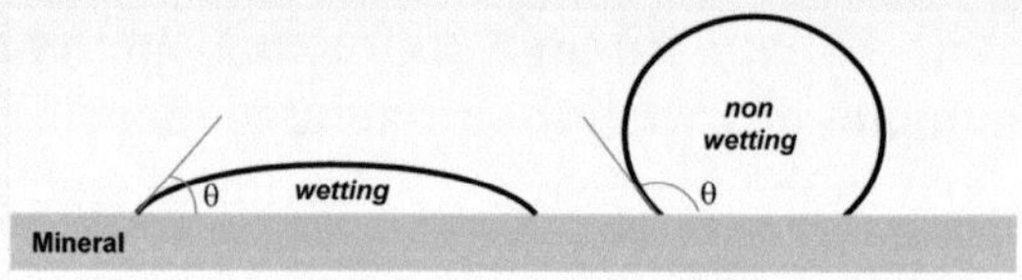

FIGURE A The geometry of the fluid droplet on the mineral surface reflects the interaction between the two fluids and the mineral substrate and can be used to determine surface tension and contact angle.

The balance between capillary forces (function of surface tension T_s and contact angle θ), viscous drag (function of fluid viscosities μ_A and μ_B), and weight or buoyancy (function of the difference in unit weight of the fluids Δγ) results in the invasion pattern of an immiscible fluid A into the rock mass saturated with a fluid B. These forces can be combined into dimensionless numbers (Lenormand et al., 1988; Pennell et al., 1996):

$$N_C = \frac{3\mu_A v}{T_s \cos\theta} \qquad N_M = \frac{\mu_A}{\mu_B} \qquad N_B = \frac{\Delta\gamma d^2}{6T_s \cos\theta}$$

These ratios vary widely (e.g., as a function of flow velocity, v, or pore size, d). When the two fluids have the same unit weight, $N_B = 0$, and invasion conditions are identified in the dimensionless N_C-N_M space shown in Figure B.

Experimental evidence (Fourar et al., 1993; Persoff and Pruess, 1995; Longino and Kueper, 1999; Akin, 2001; Rangel-German et al., 2006; Richardson et al., 2013) and complementary numerical studies relevant to fractures and fracture networks (Pruess and Tsang, 1990; Riaz and Tchelepi, 2006) are found in the literature.

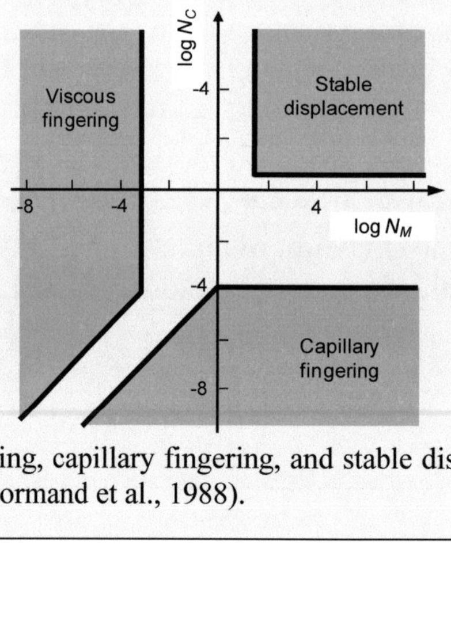

FIGURE B Space for viscous fingering, capillary fingering, and stable displacement in terms of dimensionless ratios N_M and N_C (modified from Lenormand et al., 1988).

Characterization and simulation of two-phase flow in fractured rock requires an understanding of the capillary pressure-saturation and relative permeability-saturation relationships in both the fractures and matrix. Unlike single-phase flow, the geometry of the phase boundary controls migration to a large extent. Nicholl et al. (1994) performed experiments in rough-walled fracture analogues and determined that the two-dimensional nature of the fracture (as opposed to the three-dimensional nature of the matrix and porous media) increased the importance of phase interference and trapping phenomena. The role of fracture-matrix interaction in an air-water system has been examined at a small scale (on the order of tens of centimeters) by many researchers, a summary of which can be found in Sakaki (2005). Gas-water flow in fracture networks was examined numerically by Glass et al. (2003), and the critical role of fracture intersections on the migration pathways was identified, as well as the capillary relationships. The scale of controlling phenomena in two-phase flow in fractured rock (cm) is such that explicit consideration is often not possible, and scaling or averaging approaches must be used. A mature body of work exists on scaling techniques, as well as quantification methods and parameterization approaches (e.g., Bear et al., 1993; Reynolds and Kueper, 2003).

Dissolution, Diffusion, and Migration

Constituents from an invading fluid, either LNAPL or DNAPL, dissolve into the groundwater, creating aqueous phase plumes that are then controlled by advection, dispersion, and diffusion. Dissolved constituents migrate with flowing groundwater even when the LNAPL or DNAPL source remains in place by capillary trapping (VanderKwaak and Sudicky, 1996; Yang and McCarty, 2000).

In addition to degradation reactions, contaminant mass loss to the matrix can reduce rapidly the mass of NAPLs in fractures following the cessation of migration (Parker et al., 1994, 1997; Ross and Lu, 1999; Slough et al., 1999; Reynolds and Kueper, 2002). Three snapshots during the lifespan of DNAPL in a single fracture are shown in Figure 3.7. During stage 1, continuous migration occurs while there is an active source of NAPL to feed the fracture (see Figure 3.7a). Once the source is removed, the NAPL becomes residual (stage 2) and dissolution and diffusion losses result in reduction in NAPL mass present in the fracture (see Figure 3.7b); contaminant concentrations in the matrix are near solubility levels adjacent to the fracture. During stage 3 (see Figure 3.7c) all NAPL mass has dissolved and diffused and is found in aqueous or sorbed phases in the fracture and the matrix; contaminant concentrations may decrease as clean water flows through the fracture, resulting in concentration gradients into (back) and away (forward) from the fracture. Figure 3.7 depicts contaminant concentrations migrating downward toward a higher permeability layer. Concentrations can remain above regulatory criteria for many years following the disappearance of the DNAPL phase (see Figure 3.8).

The rate of NAPL loss in the fracture depends on the solubility and diffusivity of the chemical components and the degree of sorption of the chemical(s) within the fracture and matrix. In fact, once stage 2 is reached (see Figure 3.7b), the lifespan of the contaminant as a separate phase can be on the order of days if solubility is high, fracture aperture is small, and there is an active groundwater flow regime (contaminants in NAPL form are most likely to remain in large aperture fractures with limited or no advection and low matrix porosity).

The dissolved phase storage capacity of the matrix most often exceeds greatly the NAPL-phase storage capacity of fractures. This marked difference in storage capacity could allow for the complete disappearance of the NAPL phase from fractures into the matrix. A complete change in state as the contaminant originally introduced in the NAPL phase gradually transitions to dissolved and sorbed forms would result. The rate of mass loss from fractures is such that the lifespan of the NAPL phase in fractures can be as short as several days (Parker et al. 1994; see Figure 3.9).

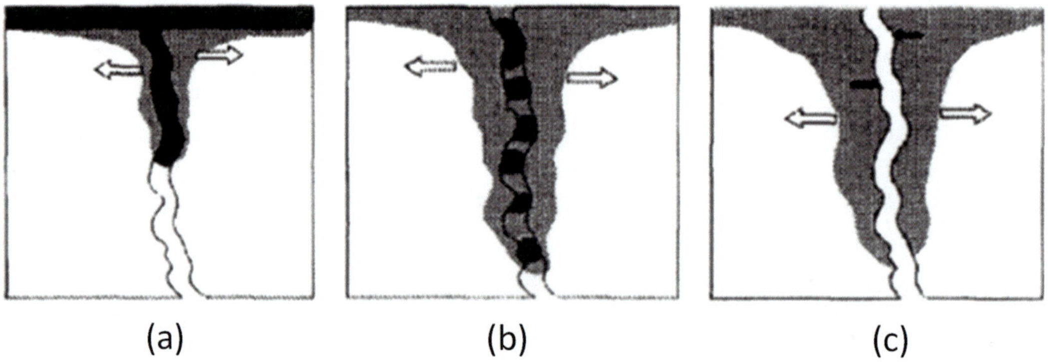

FIGURE 3.7 Stages of NAPL in a fracture. The fracture pictured represents a low permeability layer between two higher permeability layers. Stage 1 (a): there is an active source feeding invasion into the fracture. Stage 2 (b): the source has been removed; dissolution and diffusion reduce the NAPL mass in the fracture. Stage 3 (c): the NAPL mass has vanished as a separate phase and chemical concentrations in the fracture decrease as water flows through the fracture resulting in diffusion in two directions (forward, away from the fractures [white arrows] and back diffusion toward the fractures [black arrows]). SOURCE: After Parker et al., 1994.

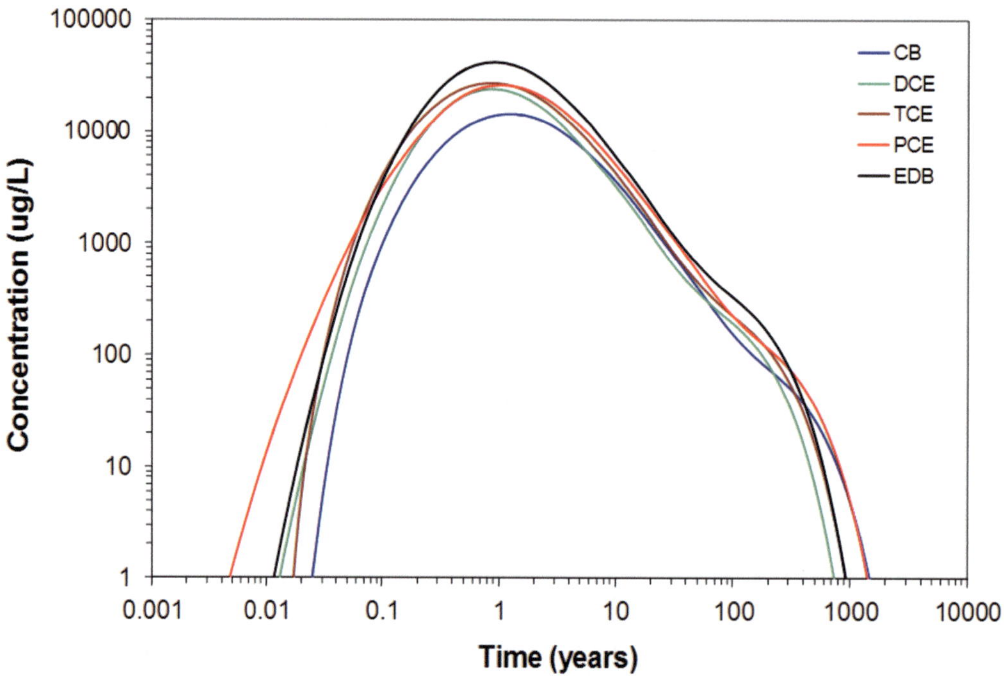

FIGURE 3.8 Concentration of contaminants exiting the fracture depicted in Figure 3.7. NOTES: CB = chlorobenzene, DCE = dichloroethene, EDB = ethylenedibromide, PCE = tetrachloroethene, TCE = trichloroethene. Matrix porosity = 0.3, fracture aperture = 35 µm, and groundwater velocity through the fracture = 0.64 m/day. Concentrations exiting the fracture are above the regulatory limits for hundreds of years longer than the pure phase (NAPL) mass is in existence. SOURCE: Reynolds and Kueper, 2002.

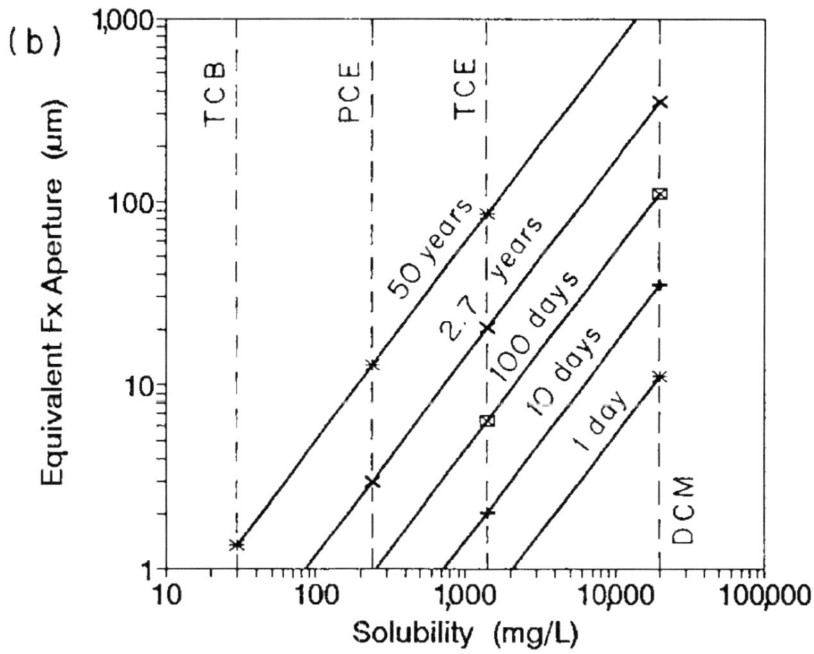

FIGURE 3.9 Lifespan of various chemicals in single fractures in typical sandstone/shale (porosity = 0.1). The solid lines represent the time required to completely transfer DNAPL from an initially completely DNAPL-filled fracture to the matrix (in dissolved phase) via dissolution and diffusion. Chart developed based on equation [4] in Parker et al. (1994), which is derived from Fick's Law and assumes a single uniform aperture fracture surrounded by an infinite matrix block.

The importance of diffusive mass loss to the matrix described here is directly proportional to the type of rock under consideration. The importance increases with increasing matrix porosity, increased fracture spacing, increased time since contaminant release to the subsurface, and decreased fracture aperture. For example, consideration of diffusive loss is critical in lightly to moderately fractured sandstones, but is likely to be less critical in highly fractured granite.

FINES MIGRATION AND ENTRAPMENT: EMERGENT TRANSPORT PROCESSES

Advective flow can mobilize and transport suspended fine mineral grains. Fines migration can lead to clogging at narrow throats along fractures and alter flow conditions. Furthermore, mobilized fines can be carriers for contaminants (e.g., colloidal transport of plutonium and water-NAPL Pickering emulsion transport).

Conditions for Fines Migration and Entrapment

Fine-grained sediments within fractures can rest on fracture surfaces or on larger grains until the balance between buoyant weight, drag forces, and electrical interactions prompt their mobilization. During migration, fines may also experience inertial forces. The relative balance between these forces is captured in dimensionless ratios such as those in Table 3.2.

TABLE 3.2 Governing Forces: Dimensionless Ratios

Name	Definition	Physical Meaning
Archimedes number	$Ar = \frac{v\mu}{d_p^2 g \Delta\rho}$	Ratio between viscous and gravitational forces
Froude number	$Fr = \frac{v^2}{d_p g}$	Ratio between inertial and gravitational forces
Particle Reynolds number	$Re = \frac{v d_p \rho_f}{\mu}$	Ratio between inertial and viscous forces
Maxwell number	$Mx = \frac{A_H}{s d_p g \Delta\rho}$	Ratio between van der Waals attraction and gravitational forces

The ratios in Table 3.2 are written in terms of particle size d_p, fluid velocity v, dynamic viscosity μ, gravity g, the Hamacker constant A_H, interparticle separation s, and both particle and fluid mass densities ρ_p and ρ_f. Fines may migrate in the rock matrix; however, the high seepage velocity in fractures compared to that in the matrix sustains preferential fines migration along fractures. Fines migration will continue as long as the grain size is significantly smaller than pore constriction thresholds (e.g., when the grain size is less than one-tenth of the pore size). Note that the migrating grain can be significantly smaller than the constriction, yet multiple fine grains form bridges at pore constrictions and stop migrating.

Pore fluid chemistry and temperature determine the surface charge and zeta potential (i.e., electrokinetic potential) of fracture surfaces and fines, and the conditions conducive to chemically induced fines migration and clogging (Khilar and Fogler, 1984; Kia et al., 1987; You et al., 2013). In particular, fines agglomeration occurs when the fluid has a high ionic concentration or a pH near the isoelectric point (i.e., no net electrical charge). In this case, small colloidal particles (particles between 1–1,000 nm in solution) group into larger flocks or against the joint surfaces. Colloidal contaminants exhibit different transport behavior than conservative solute tracers; in fact colloids may travel faster than solutes because of high velocity preferential flow paths, lack of diffusion into the matrix, size exclusion (with respect to asperities on fracture surfaces), and charge effects (Vilks et al., 1997; Becker et al., 1999; Knapp et al., 2000; McCarthy et al., 2002).

The wettability characteristics of joint surfaces (i.e., the ability of the surface to maintain contact with a fluid) may favor the release of fines. Colloidal fines held by water-wet surfaces are most likely to be released in the presence of water; conversely, fines held onto oil-wet surfaces would be released under oil saturation. Reactive fluid flow, mixed-fluid flow, and fines release-migration-entrapment may be concurrent and coupled in fracture networks, such as the release of clay fines from carbonates during acid flow for oil recovery (e.g., Sakar and Sharma, 1990; Weisbrod et al., 2002; Zhang et al., 2012).

Entrapment and clogging develops when fines become trapped at fracture throats. Grain bridging extends the opening size, O, that may clog to up to 4-6 grain diameters, d—that is O/d < 4-to-6 (Valdes and Santamarina, 2006). In the case of fine grains, flock formation extends this range even further, and the fracture constriction size may exceed 10 grain diameters and still experience clogging. Fines clogging is exacerbated by biofilm growth.

Clogging Mode and Topology

Local clogging of fractures can alter the flow field, increase the flow velocity nearby, and prompt further accumulation. The hydraulic conductivity in a fracture being clogged may eventually decrease by two or more orders of magnitude (Shin et al., 1999; Ross et al., 2001; Arnon et al., 2005). In radial flow during extraction (uniform 360 degree flow toward a single point), clogging is a self-stabilizing process and results in a clogging ring around the extraction well, but not necessarily against it (Valdes and Santamarina, 2006). On the other hand, fines in the formation are flushed away near injection wells where the flow velocity is high, as long as fracture constrictions are large enough so that granular bridges are unstable. Flushing involves positive feedback: flow localizes along flushed paths and prompts further removal of fines away from the well, leaving behind fingers without fines (see Figure 3.10).

HEAT TRANSPORT AND THERMAL PROCESSES

The fate and transport of contaminants in fractured rock alters and is affected and can be controlled by thermo-hydro-chemo-bio-mechanical processes. Previous sections explored the role of hydro-mechanical, chemical, and biological effects. This section reviews thermal effects, which have special relevance in heat-generating sources (including radioactivity and geothermal) as well as for thermal remediation strategies.

Thermal Rock Mass Properties

Thermal properties relevant to processes in fractured rock include latent heat, specific heat, thermal conductivity, and thermal diffusivity. Nominal values for fractured rock components are summarized in Table 3.3. The values in the table suggest that (1) the energy required to vaporize water is equivalent to the energy required to heat the same mass of granite, shale, or carbonate by $\Delta T > 3{,}200°C$; (2) the conductivity of rock-forming minerals is much greater than the conductivity of air and even water; and (3) the conductivity of dry fractured rock is much lower than the conductivity of the minerals the rock comprises. Thus, the local resistance to heat conduction (i.e., thermal contact impedance) hinders heat conduction in fractured rock.

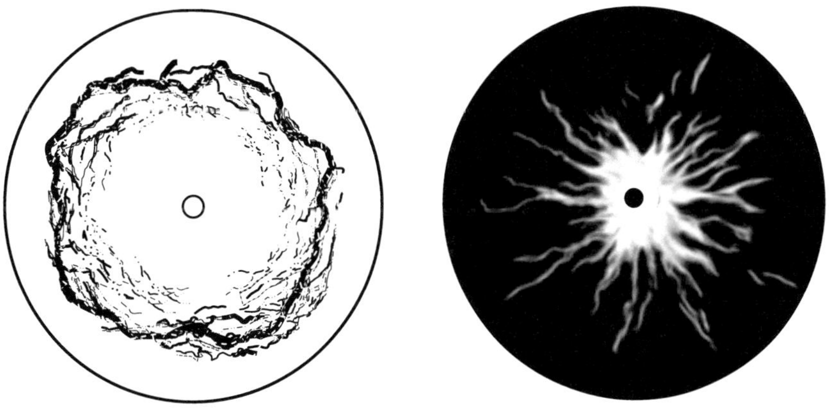

FIGURE 3.10 Fines migration in porous networks. (a) Clogging ring formed around an extraction well. (b) Erosion fingers that form around an injection well as fines are removed by high velocity fluids. SOURCE: Valdes and Santamarina, 2006.

TABLE 3.3 Nominal Values of Latent Heat, Specific Heat, and Thermal Conductivity for Fractured Rock Components

Component	Latent Heat $J \cdot g^{-1}$	Specific Heat $J \cdot g^{-1} \cdot {}^{\circ}C^{-1}$	Conductivity $W \cdot m^{-1} \cdot {}^{\circ}C^{-1}$
Air		1.01	0.02
Water	Freezing at 0°C: 334 Vaporization at 100°C: 2,260	water: 4.2 ice: 2.1	water: 0.58 ice: 2.1
LNPL (benzene)	Vaporization at 80°C: 400	1.75	0.14
Minerals and intact rock		quartz: 0.7 granite: 0.7 limestone: 0.8 shale: 0.6	quartz: 11 granite: 2–4 limestone: 1.3 shale: 1.6

NOTES: Thermal diffusivity is equal to the thermal conductivity divided by heat capacity. Conductivity and specific heat are temperature dependent. The specific heat, c, of dry and wet/saturated fractured rock masses can be readily computed as a mass-weighted combination of the specific heat of constituents. The thermal conductivity λ of dry and wet/saturated fractured rock depends on mass fractions, the thermal conductivity of individual constituents, and their spatial arrangement; in the absence of geometric characteristics, upper and lower bounds can be estimated using series and parallel configurations or the narrower Hashin-Shtrikman upper and lower bounds (Mayko et al., 2009).

The analysis of thermal conduction paths in fractured rock masses provides valuable insight for the interpretation of thermal conductivity, its variation with rock mass structure, the effects of fluids, and state of stress (see Figure 3.11). Thermal conduction pathways include

- Heat conduction within a rock block, determined by mineral composition, in particular, the fraction of quartz.
- Heat radiation across fractures into the surrounding medium (dominant at high temperatures).
- Solid-to-solid conduction that may occur through the contacts between two rock blocks (this flow path is effective stress dependent).
- Solid-fluid and fluid-solid conduction that occurs between the fluid in fractures and the rock block. This transport path is responsible for heat storage in the matrix and release back into the fluid in fractures. The presence of water at contacts has a pronounced effect on solid-solid conduction.
- Conduction in fluid. Advective fluid flow is a major contributor to heat transport. In the absence of advective currents, convection may contribute to heat transport in fractured rocks when fracture apertures exceed the millimeter scale.
- Porosity and apertures decrease and effective stress increases with depth. Solid-to-solid conduction becomes increasingly more important in deep fractured rock masses.

Heat conduction by advective fluid flow can be estimated from the fluid flow velocity v [m/s], mass density ρ [$g.m^{-3}$], heat capacity c [$J.g^{-1}.K^{-1}$], and its temperature difference with the rock mass ΔT [K]. Heat conduction through the rock matrix depends on its thermal conductivity k_T [$W.m^{-1}.K^{-1}$], and the imposed thermal gradient i_T [$K.m^{-1}$]. If the cross-section of fractures relative to the total area is α, then the flow velocity when the heat transported by the moving fluid equals the heat flux along the rock matrix is

$$v = \frac{k_T i_T}{\rho c_{fl} \Delta T} \frac{1-\alpha}{\alpha}$$

The required flow velocities computed for most field conditions are very small. Therefore, prevalent heat conduction by advective fluid flow (path 6 in Figure 3.11) is anticipated for most field situations.

Advective fluid flow along fractures combines with liquid-solid and solid-liquid conduction in and out of rock blocks to sustain effective heat transport during cyclic fluid flow with zero fluid mass flux at the end of the cycle (experimental evidence in Yun et al., 2011; path 8 in Figure 3.11). This process is analogous to chemical transport in cyclic fluid flow sketched in Figure 3.2. Extensive testing and analyses of this topic has been undertaken at the Yucca Mountain site including large-scale heat tests (e.g., Buscheck and Nitao, 1993) and modeling exercises (e.g., Birkholzer and Tsang, 2000).

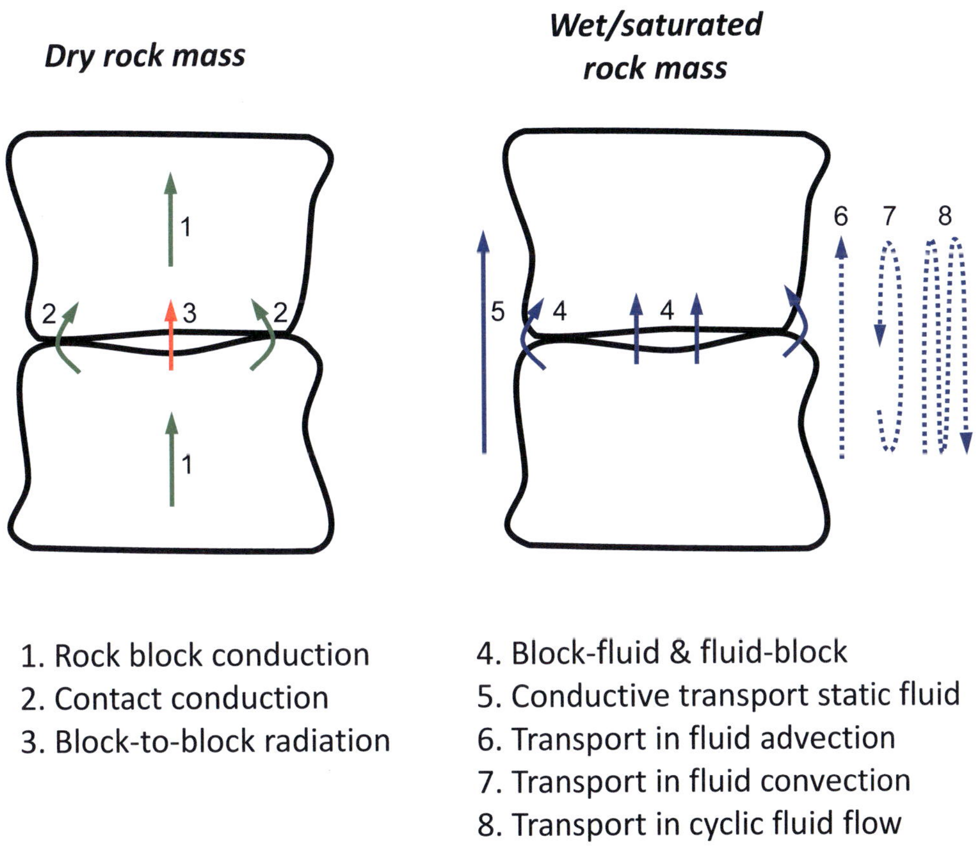

FIGURE 3.11 Heat transport pathways in fractured rock masses (refer to text for details). Heat transport from the lower to the upper block takes place within the rock block, along solid-solid contacts, and through various interactions with the surrounding fluid. Eight different pathways are labeled in the figure.

4

Conceptual and Numerical Model Development

Site models are necessary to build a conceptual picture of a site's geology and hydrology, to interpret field experiments, to estimate where and how fast contaminants may migrate, and to aid design of remediation processes and forecast performance. This chapter outlines an approach for fractured aquifer model development and emphasizes a conceptualization of the fractured rock environment that captures important features and processes of the geologic, hydrologic, and geomechanical environments and can be used to inform engineering and resource management decisions. This report will refer to such conceptual models as hydrostructural models. A site conceptual model developed early in site management and engineering can guide characterization, design, and choice of detailed numerical modeling approach appropriate for a given site, project goals, and available resources. An appropriate range of conceptual models underlies more detailed modeling efforts. Relatively simple calculation methods are available to estimate the bounds of fluid and contaminant behavior and uncertainties for even the most complex fractured rock site. This type of calculation is hereafter referred to as a scoping calculation.

Hydrostructural conceptual model development has been used by some in the European high-level radioactive waste community (e.g., Blomqvist et al., 1998) to characterize more significant fractures as deterministically as feasible. A motivation to use hydrostructural models for fractured rock is the general observation that a few discrete fractures tend to dominate flow and transport (see Box 3.1). The art of hydrostructural modeling includes distinguishing between these significant fractures and "background" fractures, which can be characterized adequately using, for example, stochastic approaches. This construct for conceptual model development is emphasized in this report because it encourages a more interdisciplinary approach to decision making related to engineering, management, or remediation of the fractured rock environment. The term emphasizes that a new way of thinking about engineering in the fractured rock environment is necessary. There remains, however, the significant challenge of how hydrostructural models and site characterization methods can contribute to the deterministic identification and interconnection of key permeable fractures. Lacking their identification, practice depends on stochastic models, which are often less satisfactory.

Conceptual models may be as simple as a pictorial representation of features, processes, and events, or can be detailed to the point at which parameters for those features, processes, and events are assigned theoretical values that can be used and tested in numerical models. This chapter describes how to conceptualize and quantify parameters of the fractured rock site conceptual model and the utility of doing so. The discussion concludes with an outline of methods used to construct and analyze numerical models to ensure there are quantitative connections between observations and numerical models. Simple approaches and quantifying both conceptual and parameter uncertainty are emphasized. This chapter focuses on approaches to numerical modeling rather than specific programming packages. Several commercial and non-commercial numerical models are available. Chapter 5 describes methods to collect data to support detailed numerical models.

DEFINING AND DEVELOPING HYDROSTRUCTURAL MODELS

Hydrostructural models for fractured rock, like numerical models discussed later, are approximations of reality used to describe the geologic and hydraulic characteristics of a site. Hydrostructural models are site specific, and usually include some combination of discrete features (i.e., fractures, faults, karst, dikes, brecciated formations, and permeable bed boundaries) and rock matrix blocks (defined by the rock material between discrete features). Without an appropriate underlying hydrostructural model as a conceptual framework, analysis or numerical modeling will not yield a thorough understanding of discrete fracture geometry and its effect on flow, transport, and storage for even the smallest project.

The term "upscaling" is used to describe the process of going from a detailed, high-resolution model to a coarse model, for example, through averaging (e.g., Rubin and Gómez-Hernández, 1990; Durlofsky, 1992; Ouenes and Hartley, 2000). Upscaling is often necessary to allow numerical computation at larger scales with fewer computational resources (e.g., Durlofsky, 2002). It is required, generally, to translate from a hydrostructural to a numerical model. Hydrostructural models, therefore, can be simple, but need to represent the full range of geologic features, characteristics, and geometries that are important hydraulically to the processes of interest, so that those processes, features, and behaviors are not lost during upscaling.

Not all hydrostructural model elements need to be quantified, and models should be informed but not limited by existing data. Sparse, biased, or censored data may result in an erroneous conceptual model. A vertical borehole, for example, may not capture hydraulically important vertical fractures known to exist at a site. It is therefore better to include those vertical fractures in the hydrostructural model, even if their parameters are unknown, rather than omitting them because site-specific data are not (and may never be) available. Hydrostructural models are not constrained by the same assumptions as specific numerical models (e.g., gridding and grid resolution, equivalent continuum concepts, or simplified porosities).

Hydrostructural Model Templates

It has been concluded, based on limited empirical data, that analysts often have difficulty selecting an appropriate conceptual model and that initial models chosen are often found to be incorrect (Bredehoeft, 2005). Hydrostructural conceptualization begins with the simplest visualization of site geology and grows in complexity as information is gathered until transport and remediation processes can be incorporated into the model. A dolomite site may be conceptualized, for example, by first assuming a set of parallel bedding planes and two sets of fractures perpendicular to bedding. Detail is added to the model to reflect solution features likely found in dolomite. More details are added as fracture and solution feature properties are characterized. The hydrostructural model should eventually describe the roles in and relative importance of fractures and rock matrix in flow (permeability; see Box 4.1), storage (porosity, storativity), and connectivity (discrete pathways). Table 4.1 provides generalized hydrostructural model examples for some geologic settings that may be good starting points for conceptualization. The maturing hydrostructural model can inform characterization strategies (see Chapter 5), data from which can validate and inform the hydrostructural model (see Chapter 7 for information about the use of the observational method to update models).

Representing a fractured rock as a system of discrete fractures and rock matrix is critical to conceptualizing contaminant transport and remediation. The conceptualized division is not strictly consistent with physical reality, however, because fractures can occur at scales from microns to kilometers. The scale of the contaminant plume being studied, for example, may determine which fractures and fracture characteristics (e.g., porosities and fracture permeabilities) are relevant to transport or storage, and therefore whether they should be conceptualized as part of the effective rock matrix or fractures. The full range of storage and flow porosities needs to be considered and assigned in light of the information needed for forecast of contaminant transport and design of a remediation process.

BOX 4.1 Contributions of Fractures and Matrix to Flow and Storage

Rock fractures and matrix both influence permeability, connectivity, and storage porosity in fractured rock environments. Several approaches have been developed to help practitioners evaluate how best to analyze a given fractured rock. These include the classic formulation proposed by Nelson (2001) that defines fractured rock in terms of the relative contributions of fractures and matrix to porosity and permeability. Nelson's approach is valuable, but it does not take into account the defining role of fractures in connectivity—neither in the control of solute transport pathways nor in the exchange of fluids and contaminant between fractures and matrix.

The figure below expands on the roles of rock matrix and fracture porosities posited by Nelson by considering explicitly the effect of connectivity. Fluid and contaminant exchange between fractures and matrix need to be accounted for where matrix porosity is significant, even when fractures are modelled explicitly. Connectivity in the figure is considered both in the positive sense (e.g., preferential flow pathways and exchange between fractures and matrix) and in the negative sense (e.g., barriers to flow such as sealing and partially sealing faults).

In rocks in which fracture effects on connectivity are significant (i.e., Type I and Type II rocks in the figure), explicit accounting of connectivity of the fracture network is necessary even if total fracture porosity—and even total fracture permeability—is less than that of the rock matrix. Where rock matrix storage is significant (i.e., Type II and Type IV in the figure), timescales of fracture/matrix fluid exchange need to be considered in any evaluation of transport and storage. Significant fracture storage may occur where there is no significant fracture connectivity (i.e., Type III rocks). This is possible, for example, in rocks with no fracture network but with localized karst zones. This approach forces the practitioner to consider explicitly how elements of the rock mass system are connected and where storage occurs in the system. This is important even when a single porosity or dual interacting continuum modeling approach is adopted.

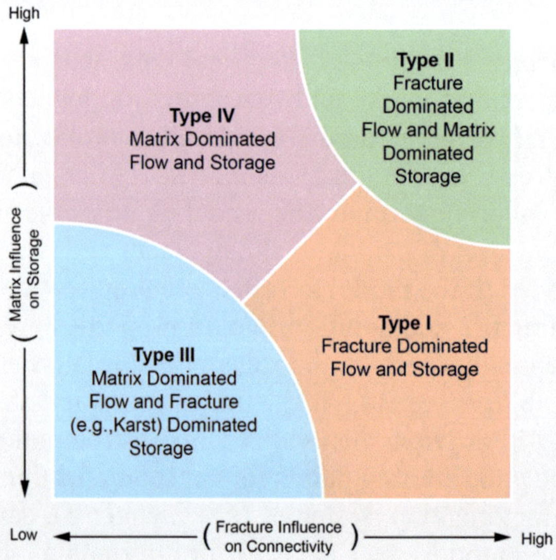

FIGURE Conceptual taxonomy of flow regimes in fractured rock aquifers based on contributions of fractures and matrix to flow and storage.

TABLE 4.1 Sample Hydrostructural Model Templates for Generic Geologic Settings[a]

Geologic Setting	Starting Hydrostructural Model Template
Fractured crystalline rock (e.g., granite, diorite) 	Very low (< 1%) rock matrix porosity, with fractures at a continuous range of scales from mm to 10s of km providing. Fractures occur in one or more sets, with both lateral and vertical spatial variation. Rock mass porosity and permeability based on smaller fractures and rock matrix. Essentially no flow and very little storage in matrix. (Photo of Sentinel Ridge. Courtesy of L. Lau.)
Fractured carbonate rocks (e.g., limestone, dolomite, oil shales) 	Mechanical unit bedding with fractures perpendicular to bedding. Because bedding can be distorted by folding or faulting, fractures will be oriented in a local coordinate system consistent with bedding. Fractures will be bound by mechanical units. There are frequently solution enhanced features (karst), including thin brecciated beds. Superposed on this may be fracture sets related to faulting (in a global coordinate system) and folding (in a bed-local coordinate system). Rock matrix porosity and permeability can be significant. (Photo of part of the Supai Group in the Grand Canyon. Courtesy of the U.S. Geological Survey.)
Argillaceous rocks (e.g., shales) 	Similar to fractured carbonates, with both bed-local coordinate system fractures, and superposed tectonic and structural fractures. Rock matrix permeability is generally low enough to inhibit flow such that smaller fractures may form a major portion of the rock mass permeability and porosity. (Photo of the Geneseo/Burket gas shale at Taughannock Falls State Park, Trumansburg, New York. Natural hydraulic factures emerge from the top of the black gas shale and penetrate upward into a gray shale. Courtesy of T. Engelder, The Pennsylvania State University.)
Extrusive Igneous rocks (e.g., tuff, basalt) 	Unique in their complexity. Fracture patterns can be systematic (particularly in basalt) but can exist in layering of varying porosities made complex by cooling processes, multiple eruption and faulting events, igneous intrusions, brittleness, and interbedding of paleo-alluvial and ash-flow materials. (Photo of pillow basalts at Mary's Peak, Oregon. Courtesy of R. Keller.)

[a] This table does not represent all geologic environments and should not be considered a comprehensive tool.

Microstructural Models of Fractures

Fractures are often not defined by a single pair of parallel flat surfaces as they are often modelled. Because diffusion and surface sorption rates can be controlled by the ratio of flowing volumes to fracture surface area, diffusion and sorption rates are a function of the geometry and topology of surfaces within a fracture. Individual fractures may therefore need to be characterized at smaller scales—even down to the micro level: a microstructural model. A microstructural model allows conceptualization for multiple fracture surfaces, infillings, coatings, and altered rock zones that make up a fracture. It is these characteristics that determine the geochemistry of retention and the relationships between mechanical, filtering, transport, and hydraulic apertures. See Box 4.2 for an example of a microstructural model.

Hydrostructural Conceptualization of Faults

Details about faults that control fluid flow and contaminant fate and transport can also be incorporated into hydrostructural models. Only a limited number of fractures, some of which may be faults, need to be considered explicitly in hydrostructural modeling because some fractures do not conduct fluid or accommodate storage or transport. For the purpose of conceptualization, fractures can be grouped as

- Low transmissivity fractures that add to pore volume but do not conduct flow. Fluid behavior in these fractures is similar to fluid behavior in the surrounding rock matrix, and the fractures can therefore be conceptualized as part of the effective rock mass. Transmissivity considered "low" depends on the application—for radioactive waste, this value might be 1×10^{-8} m^2/s, whereas it might be 1×10^{-7} m^2/s for oil.
- Higher transmissivity fractures that provide significant flow and reactive surface area.
- Significant conductive and flow barrier features defined explicitly to describe flow pathways and rock matrix storage.

The fault itself can be modeled as a fault core and a damage zone. Each may be conductive or not, and the damage zone can be small or large compared to the fault core. Elements incorporated into the hydrostructural model include

- Fault core gouge—ground material found between fault walls formed as result of mechanical erosion resulting from fault displacement. Fault core gauge can be several meters wide and create barriers to flow across a fault.
- Fault core breccia and other high-permeability and high-porosity infill materials that can provide major storage for contaminants or key flow pathways in the fault core.
- Hangingwall and footwall disturbed zones. These zones can have significant permeability and porosity parallel to faults, even if the fault itself is a flow barrier (low permeability) perpendicular to the fault due to core gouge.

BOX 4.2 A Microstructural Model by the Swedish Nuclear Fuel and Waste Management Company

The figure below is a microstructural model developed by the Swedish Nuclear Fuel and Waste Management Company (SKB) (Winberg et al., 2003) and represents advection dispersion and diffusion processes and flow paths at the micro scale. Characteristics of specific features in the SKB investigations were quantified to refine the hydrostructural model. Because matrix diffusion rates can increase significantly in the presence of multiple surfaces, attempts were made to quantify the number of pairs of fracture surfaces, particularly for larger features such as faults. Similarly, because the mix of geologic materials on any given transport pathway contributes to the uncertainty in transport times, different rock matrix materials, each with its own porosity, tortuosity, and geochemical properties (i.e., sorption rates), were considered (e.g., Andersson et al., 2002). Empirical measures of complexity were used to help to quantitatively evaluate the storage and transport capacities of specific fracture pathways. Additionally, variations in fracture orientation can vary significantly between different borehole intersections of the same fractures—orientations in one borehole are not necessary indicative of the average fracture orientation at larger scales.

Specific features on site were mapped and quantified using, for example, borehole television and corelogs (e.g., Andersson et al., 2002). With such parameterized information derived from site characterization, the number of conductive fractures was determined, and microstructural models were refined in greater detail. Information at this level of detail is important when developing a hydrostructural model because

a. advective transport velocities depend on the distribution of flow channels within rough and mineralized fracture planes;
b. dispersion and mixing is significantly influenced by fracture intersections—even those within fractures that can be simplified at larger scales as single planes;
c. reactive surface area controls sorption and diffusion processes, and the amount of available surface area for a given pathway is controlled by fracture complexity;
d. the variation of porosity and mineral geochemistry effects solute retention processes; and
e. the size and shape of rock matrix affect the rate at which hydraulic and geochemical equilibrium is reached. Slab-shaped (long, narrow) rock matrix blocks may reach geochemical equilibrium faster than an equivalent volume of spherical (i.e., equi-dimensional) rock matrix blocks.

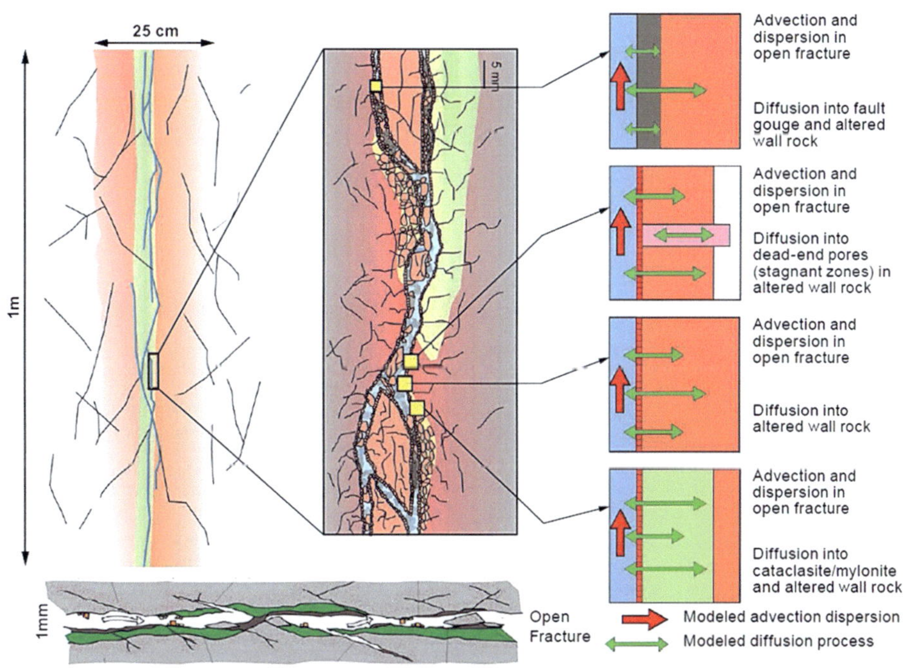

FIGURE Microstructural model approach showing progressively more detailed structures and eventually the predominant flow path and transport and storage processes at micro scale. SOURCE: Courtesy of SKB (adapted from Winberg et al., 2003).

Figure 4.1 illustrates the range of fault-zone architectures and associated permeable structures possible in a hydrostructural model. Each corner of the figure represents an end member along a continuum of possible scenarios. When the damage zone is small, and the fault core is also a small fraction of the local volume, the fault itself can be a conductive feature, with fluids traveling along the fault. If the damage zone is a region that contains many fractures, and the fault core is relatively small, then both the fault core and damage zone contribute to flow in a distributed conduit. If the fault core is large and acts as a flow barrier, and the damage zone is also large, then the conduit created by the fault can allow enhanced flow along the fault but block flow across it. The combination of a small damage zone with a fault core barrier can create a barrier to flow transport across the fault and not enhance flow along the fault.

These end member examples represent very different settings for flow and exchange of contaminants with nearby zones. Failure to account for these elements in the hydrostructural model can limit the usefulness of the models for design and forecast of contaminant transport and mitigation.

Rock Matrix Conceptualization

There are at least two timescales for remediation: one associated with groundwater residence times and another associated with the diffusion of contaminant in the immobile porosity of the rock. This is true for single-porosity porous (non-fractured) and multiple-porosity fractured rock. Blocks of rock matrix may be of different porosity than the fractures surrounding them, and the shape and size of the matrix blocks affect diffusive, advective, geochemical, and other types of exchanges between fracture and matrix. The oil industry conceptualizes interactions between rock matrix and fractures in terms of three alternative rock matrix block aspect ratios: slab, matchstick, and sugar cube (or "sphere"; see Figure 4.2). Each of these model blocks exhibits distinct contaminant transfer behavior between the advective (fracture) porosity and the immobile and storage (rock matrix) porosity, affecting both hydraulic pressures and geochemical equilibrium. The timescales of transfer depend on the characteristic size defined by the minimum rock block dimension (Kazemi and Gilman, 1993; Carrera et al., 1998). In remediation, contaminants in matchstick-shaped blocks are accessed quickly, while those in a matrix with sugar cube dimensions are accessed more slowly. The shape of the tail of the contaminant plumes is therefore influenced by matrix block shape, and block size also influences contaminant storage.

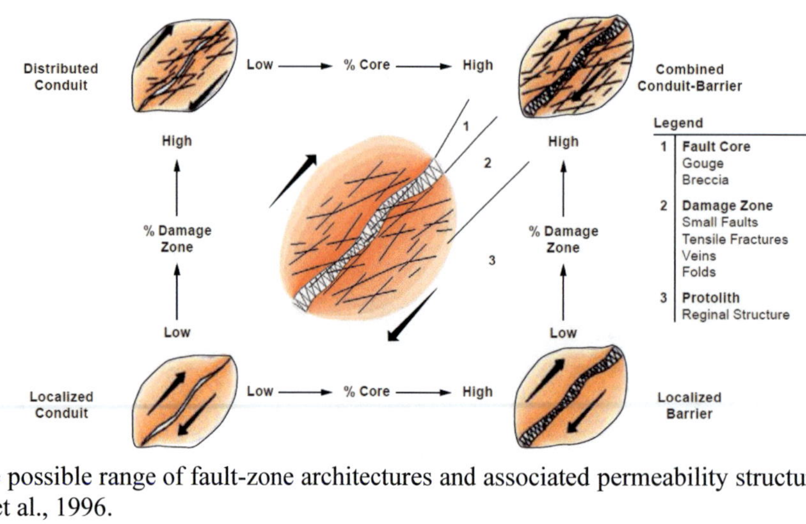

FIGURE 4.1 The possible range of fault-zone architectures and associated permeability structures. SOURCE: Modified from Caine et al., 1996.

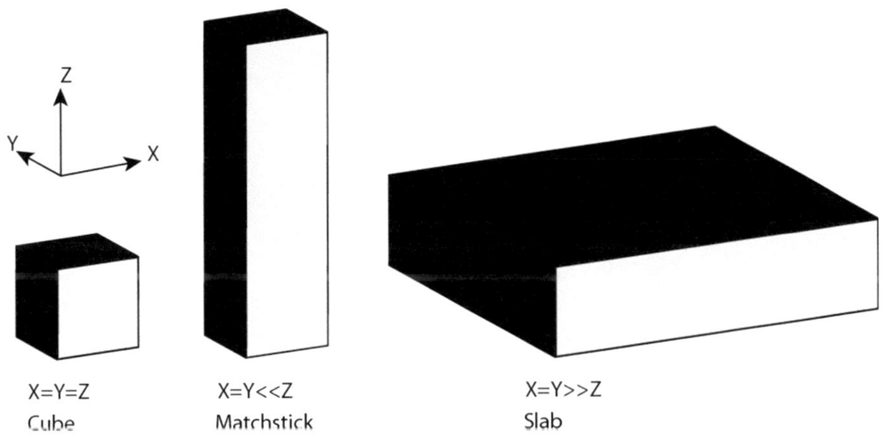

FIGURE 4.2 Oil industry conceptualization of rock matrix storage porosity in terms of block (sugar cube), matchstick, or slab shapes. SOURCE: Fox et al., 2009.

Matrix blocks are commonly modeled as sugar cubes, but better models result if variability is accounted for. The distribution of rock block shapes in the hydrostructural model can be calculated from estimates of the storage and flow porosities and the estimated spatial distribution of contaminants.

QUANTIFYING THE HYDROSTRUCTURAL MODEL

Site characterization data will inform but not completely constrain the hydrostructural model. The next stage of model development is to add parameters that describe each of the fracture and matrix elements.

Data Integration

Generic parameterizations applied early in hydrologic modeling avoid common bias in the models. Values derived from field tests are not always representative across a site. Every field test or measurement has some spatial scale of investigation (see Chapter 5), and results of those tests should be applied to other length scales only with caution. For example, a single hydraulic test might yield a specific local transmissivity, but this value may or may not be representative of transmissivity on larger scales or in adjacent areas. Such measurements can do little to constrain the range of values that may be present at the site scale. Initial parameter value estimates can be based on conditions observed at similar sites.

Not all elements of a site hydrostructural model can be parameterized deterministically. Successful parameterization depends on the ability to integrate multiple kinds of information and lines of evidence from multiple sources (see Chapter 5 for discussion of data integration). Better predictions and practices are possible if both parametric uncertainties and uncertainties related to the appropriateness of the hydrostructural conceptualization are considered. Table 4.2 lists the types of data that could be integrated to determine various hydrostructural elements associated with fractures and rock matrix. Hydrostructural elements related to boundary conditions do not appear on the table because boundary conditions do not conveniently fit into a table of this type. The next sections describe more fully the parameterization of fractures, rock matrix, and boundary conditions.

TABLE 4.2 Data Integration for Parameterization of Hydrostructural Models

Method of Parameter Determination	Hydrostructural Element Related to Fractures	Hydrostructural Element Related to Rock Matrix
Geologic (including core, outcrops, and lab testing)	Spatial pattern sets and orientation distributions Size, shape, roughness, and infilling Microstructural model Aperture	Shape Permeability and porosity Geochemical properties
Geophysical (including fracture image, wireline, and surface seismic methods)	Spatial pattern Transmissivity distributions Orientation distributions Anisotropy and heterogeneity Major structures (i.e., large faults, fold crests and hinge zones, dikes, large karst features)	Elastic properties
Hydrodynamic (including well- and multi-well-based tests, long-term pressure and temperature monitoring, and hydrophysical flow logging)	Transmissivity, storativity, aperture measures, porosity measures Functional relationships among hydrodynamic properties Functional relationships between hydrodynamic property and geometric properties	Hydraulic conductivity Porosities measures
Geomechanical	Strength Shear and normal stiffness In situ stress	Strength Deformability In situ stress
Thermal	Thermal deformability Functional relationships between hydrodynamic properties and temperature Heat capacity	Thermal deformability Functional relationships between hydrodynamic property and temperature Heat capacity
Geochemical	Porosity measures Diffusion measures Sorption/retention measures	Porosity measures Diffusion measures Sorption/retention measures Fluid/rock interactions: dissolution/precipitation reactions

Fracture Parameterization

Fracture parameterization within hydrostructural models is most useful for many applications if based on site microstructural models, spatial patterns, hydrodynamic properties, and geochemical and biologic properties (e.g., retention, natural attenuation, and those related to mitigation). Table 4.3 provides a list of various fracture properties and their measures. Following are examples of how various fracture elements could be parameterized.

TABLE 4.3 Fracture Measures

Dimension of Sampling Region		Dimension of Feature: Number of Fractures (0)	Fracture Trace Length (1)	Fracture Area (2)	Fracture Volume (3)	
	0	P_{00} [-] Number of fracture samples per point sample of rock mass				Point Measures
Line (Borehole)	1	P_{10} [1/m] Number of fractures per unit length of scanline *(frequency or linear density)*	P_{11} [m/m] Total fracture aperture per unit length of scanline *(lineal porosity)*			Linear Measures
Area (Traceplane)	2	P_{20} [1/m²] Number of trace centers per unit area of sampling surface *(areal density or trace density)*	P_{21} [m/m²] Length of fracture traces per unit area of sampling surface *(areal intensity or trace intensity)*	P_{22} [m²/m²] Area of fractures per unit area of sampling plane *(areal porosity)*		Areal Measures
Volume	3	P_{30} [1/m³] Number of fracture centers per unit volume of rock mass *(volumetric density)*		P_{32} [m²/m³] Area of fractures per unit volume of rock mass *(volumetric intensity)*	P_{33} [m³/m³] Volume of fractures per unit volume of rock mass *(volumetric porosity)*	Volumetric Measures
		Density		Intensity	Porosity	

SOURCE: Courtesy of W.S. Dershowitz.

Hydrodynamic Properties

Hydrodynamic properties that dominate contaminant migration rates need to be parameterized. These include transmissivity, aperture measures, storativity, and hydraulic diffusivity (transmissivity/storativity). Transmissivity, storativity, and hydraulic diffusivity should be quantified and applied considering the correlations between them and the importance of those parameters for understanding flow and transport behavior. Unless there is evidence of little variability in fracture geometry and in hydrodynamic properties, assume there will be spatial and statistical variability of these within and among fractures.

Fracture Intensity

As discussed in previous chapters, fracture intensity is the number of fractures in a unit volume. The number of fractures indicates matrix diffusion rates associated with surface fractures. Table 4.3 lists measures of fracture intensity.

Fracture Roughness and Infilling

Fracture roughness and infilling are often neglected fracture characteristics. Under any given set of boundary conditions, contaminant travel time, retention, and reactive surface areas depend on whether flow is effectively one dimensional (e.g., a pipe), two dimensional (e.g., a plate), three dimensional (e.g., a volume), or some intermediate (fractional) dimension. The nature of flow and transport dimension can be influenced strongly by roughness and mineralization, particularly at fracture intersections, and can also be a function of both fluid pressure and in situ stress magnitude and orientation.

Relative Permeability Curves

Relative permeability curves (also referred to as characteristic curves) describe permeability to individual fluids as a function of the phase saturation. They are used to parameterize multiphase flow and transport behavior. Distinct and appropriate relative permeability curves are dependent on multiple geologic and geomechanical properties (e.g., fracture roughness, infilling, and persistence). Multiphase relationships need to be described frequently through correlation and probabilistic relationships with other fracture characteristics. Caution is necessary, however, when using relative permeability relationships because the use of such pressure-saturation relationships in fractures poses challenges. These relationships were developed for continuum representation of porous media and are not necessarily directly relevant to discrete features such as fractures.

Rock Matrix Parameterization

Rock matrix parameterization is dependent on determining which fractures should be considered part of the effective rock mass and the matrix block geometries for fracture-rock matrix interaction calculations. Properties such as permeability (and its function of phase saturation), total porosity, and pore size distributions for intact rock can be characterized generally through laboratory and other testing methods (see Chapter 5). The model needs to define matrix porosity and pore size distributions to evaluate mobile and immobile contaminant transport, and mitigation and remediation activity designs. As discussed in Chapter 3, pore size is important in filtering and biological processes. Anisotropy of rock matrix parameters should always be assumed and at least approximately quantified based on available geologic and geophysical data.

Boundary Condition Parameterization

Hydrostructural model boundary conditions describe hydrogeologic features such as water levels, faults, end of aquifers, and recharge or discharge zones that affect local hydraulic conditions. Boundary conditions describe the characteristics assigned to the boundaries or edges of a numerical model that represent regional hydraulic behaviors, as well as the parameters assigned at the beginning of a numerical model test to represent background conditions. Flow, pressure, and concentration boundary conditions are important parts of a hydrostructural model—particularly as they help define dominant discrete features. Standard techniques for defining boundary conditions in hydrogeology (Bear, 1972) also are applicable to fractured rock. Boundary conditions, however, can be discontinuous. For example, a fracture connected to a fault may provide a constant head boundary condition, while an adjacent fracture connected to a pumping well may provide a flow rate boundary condition. The details of the hydrostructual model and its connectivity thus create a spatially varying hydraulic distance to defined boundary conditions. Sensitivity studies can be used to quantify the effect of alternative boundary condition definitions. In many cases, larger discrete features may themselves be the boundary conditions—for example, a sealing fault may be a no-flow boundary condition.

Boundary condition development is complex, and there is much literature on the topic (e.g., Anderson and Woessner, 1992; Barnett et al., 2012). Initial and boundary conditions—including some combination of pressure, flow, temperature, geochemical, and biological boundary states—affect model predictions and design simulations strongly. Consequently, the level of effort for establishing appropriate boundary conditions should be comparable to that used for assessing material properties.

Boundary conditions to be considered include

- large, conductive faults that act like constant pressure boundaries;
- flow barrier faults that act like no-flow boundary conditions;
- recharge and discharge boundaries identified from geochemical signatures (e.g., highest total dissolved solids at discharge);
- topographic surface water divides (used in continuum models, but are not frequently groundwater divides in fractured and compartmentalized rock systems); and
- very low fracture intensity zones that may be modeled as no-fracture-flow boundaries.

SCOPING CALCULATIONS TO ASSESS AND REFINE MODELS AND UNCERTAINTIES

Analyses of contaminant mitigation and remediation engineering solutions need not be complicated—even at complex fractured rock sites. In some cases, a simpler analysis is more helpful than the results of a more complex regional groundwater model. It is worthwhile and cost effective to use simple scoping calculations to determine the level of analysis necessary for a given engineering problem. At sites where there is evidence that a single discrete feature dominates transport, an approximate analysis to identify and quantify the pathway flow and storage properties can forecast transport accurately. At other sites, the most important process may be the exchange of contaminant between mobile fracture porosities and immobile matrix porosities. In those cases, an analytical solution of mass exchange might provide an excellent indication of the timescales of site remediation. Many important questions related to fractured rock characterization and monitoring can be expressed in terms of the probability that a given array of boreholes will intersect features of a given geometry. Equations of stochastic geometry and geometric probability (e.g., Chiu et al., 2013) can answer such questions without groundwater flow or transport modeling.

A good analysis approach is to determine first the questions to answer, then to develop efficient and effective ways to answer them using the site hydrostructural model and knowledge about available simple and complex analysis tools. Simplified analyses can determine the most sensitive assumptions and parameters in the system so that the most resources can be devoted to quantify them. Other simple analyses can determine which porosities, geometric issues, and processes to evaluate. These include geometric issues such as probabilities of intersection and the probability of connections between two points in space (i.e., between plume and discharge locations). Simplified analyses can also indicate which hydrostructural and microstructural model features may be upscaled and the consequences of upscaling. Simple analyses can be used to determine the range of results expected from more complex and expensive modeling, and whether the resulting reduced uncertainties justify the additional modeling costs.

ANALYSIS TOOLS TO INFORM MODELING

Numerous commercial and public analysis tools are available to facilitate fractured rock modeling, but it is beyond the scope of this report to assess all of these tools. A list of publicly available analytical tools has been produced by the Integrated Groundwater Modeling Center at the Colorado School of Mines.[1] The choice of tool warrants careful consideration by the practitioner, and it behooves the engineer to understand the modeling approaches incorporated into software design. Geometric calculation tools should be used to answer geometry-related questions (e.g., determining matrix block diffusion distances, or the probability that a discrete fracture pathway exists between a contaminant source and compliance boundary). Solute transport calculation tools should include the ability to analyze matrix diffusion processes, surface sorption, and rock matrix sorption without representing fractured rock as a single porosity system. Well-test interpretation tools should determine the flow dimension effects of fracture network connectivity.

Correlations and Parameterizations

The assumption of independence between fracture geometric properties (e.g., fracture size, orientation, and depth) and hydrodynamic properties (e.g., transmissivity, storativity, aperture) can have significant impact on forcasts made with such models. Independence should only be assumed where there is clear evidence that geometry and hydrodynamic properties are independent. More research is needed to better define the functional relationships between fracture geometric and hydrodyanmic properties. In the absense of site-specific evidence to the contrary, it is better to assume a correlation than to assume there is no functional relationship. Klimczak et al. (2010) and Hjerne et al. (2010) provide example correlations between fracture size and flow and transport properties.

It is also important to emphasize that fracture intensity should not be expected to show clear correlation to effective rock mass permeability (see, for example, research at the Apache Leap Research Site in Arizona; Chen et al., 2000). This is due to the wide variability in fracture transmissivity (generally many orders of magnitude); a few high-conductivity fractures contribute more to rock mass hydraulic response than a much larger intensity of less transmissive fractures. Hypothesis testing approaches to determine whether such functional relationships can be expressed in terms of in situ stress and geomechanical properties can be useful. Available correlations allow estimation of fracture transport parameters (e.g., transport aperture, transmissivity, and storativity) from geologic and geometric measurements (see Figure 4.3).

Pathway Geometry and Topology

Many fractured rock contaminant hydrogeology issues can be thought of as topological (connectivity) rather than as hydrodynamic problems. As such, they can be addressed through analysis of geometry. Geometry and topology can answer questions such as the probability that a given monitoring well configuration will detect contaminants, about whether an injection/withdrawal pair will have sufficient reactive surface area to address a certain volume of contaminated rock, or whether contaminant flowing through a fracture network in an aquitard layer will transport contaminant to an aquifer.

---

[1] See http://igwmc.mines.edu/software/category_list.html#STA (accessed December 11, 2013).

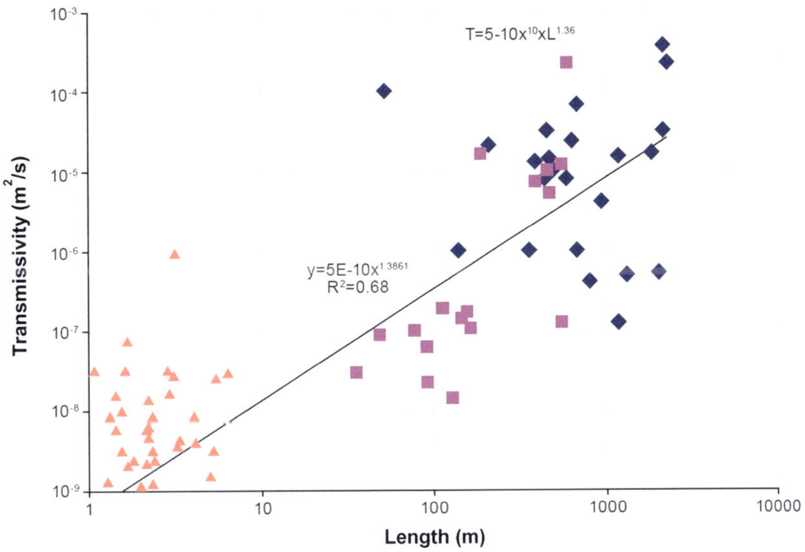

FIGURE 4.3 Correlation between fracture size and transmissivity as determined from the Äspö Hard Rock Laboratory in Sweden. Symbols represent data from experiments conducted at different scales. SOURCE: Modified from Dershowitz et al., 2003.

Scoping calculations based on graph theory analysis of the geometry of conductive features and flow barriers (e.g., Bodin et al., 2007) can address problems of topology and the probability of connections among features (i.e., Sahimi, 1994). If there is concern about the probability of rapid breakthrough at a well 100 meters from a contaminated location through a single high-transmissivity fracture, then the fracture size (radius) distribution can be used to calculate the probability of breakthrough for a single fracture of that size or greater. Such approaches allow transport prediction and remediation and monitoring design to be carried out with simpler, one- or two-dimensional methods.

Immobile Zone Interactions

Mass transfer rates and storage volumes for mobile and immobile porosities can be estimated with one-dimensional approximate calculation approaches. The approaches can therefore be used to better estimate where contaminants are likely stored, to estimate the timescales of various remediation processes, and to assess whether more complex and resource-intensive modeling and analysis is warranted (e.g., single porosity/single permeability, single permeability/dual porosity, dual porosity/dual permeability, or discrete fracture network [DFN] approaches).

Sensitivity to Coupled Mechanisms

Coupled processes can have a significant effect on flow and transport. This has been a particularly important issue in the petroleum industry, where fracture closure in response to decreased reservoir pressure with production has been found to reduce flow rates and oil production (e.g., Hansford and Fisher, 2009). Baseline scoping calculations need to account for such coupling—at least in a generic way—where significant changes over time in temperatures, pressures, or chemistry occur. Failure to include these correlations in analyses can degrade the ability to forcast flow and transport response over time, increasing both error and uncertainty.

However, coupled process calculations can complicate analyses, and they require more time or effort available for a particular application. In such cases, as a minimum, the effect of ignoring these couplings needs to be estimated. Better analysis will result when simplified scoping calculations are used to estimate the sensitivity of contaminant transport, retention, and remediation in fractured rock to coupled thermal, hydrodynamic, mechanical, biological, and geochemical processes. Estimates of the change in storativity and transmissivity as a function of stress can be used to explore fracture stiffness-aperture-transmissivity relationships. Equations such as those provided by Jiang et al. (2009) can be used. Revised storativity and stiffness values can then be applied in flow and contaminant transport calculations to estimate responses of flow and transport to changes in effective stress magnitude and orientation.

Type-Curve Derivative Approaches

The theoretical response of models to constant-pressure and other tests can be calculated, and the expression of the results can be used for practical applications. For example, calculation results can be generalized on dimensionless graphs (e.g., dimensionless pressure versus dimensionless time) for type curve analysis (e.g., Gringarten, 1987; Barker, 1988) and used to assess fracture network connectivity and dual-porosity behaviors in a simple, practical manner. Derivative approaches using type curves to analyze dynamic flow and transport data (e.g., Black et al., 1987; Doe, 1991; Acuna et al., 1995; Illman and Neuman, 2000, 2001) are powerful and underused tools. They can be used to examine fracture network connectivity and boundary condition issues that, in turn, control the flow system, interaction between mobile and immobile porosities, and boundary conditions such as flow barrier faults and connected high-porosity aquifers.

When dynamic flow or transport information is available, type-curve derivative analysis can provide significant insights. Derivative analysis and multi-rate type curve analysis both can be used to better understand flow geometry and to help constrain fracture hydrodynamic properties (transmissivity and storativity), boundary conditions, and dual-porosity or dual-permeability behavior. This approach can be used to determine when contaminant transport can be modeled as a single porosity, pipe (linear) flow process, a dual porosity (fracture and matrix) system, or a radial (confined aquifer; Doe, 1991) system. The relative importance of geometry (e.g., discrete flow channels) and matrix interactions can be assessed by reviewing hydrodynamic data using flow dimension and derivative approaches. Better forecast of contaminant transport, storage, monitoring, and remediation will be possible with firm understanding of dimension of flow and the effect of flow boundaries.

Simplified Transport Approaches

There are several multiple-immobile-zone solute transport approaches that provide quick estimates of solute transport, even in complex multiple-porosity systems and can therefore be used in sensitivity studies to examine assumptions, to prioritize data collection, and to scope methods for mitigation. These include Laplace Transform Galerkin methods (Sudicky and McLaren, 1992), Lagrangian methods (Cvetkovic and Dagan, 1994), and continuous time random walk analysis approaches (Bijeljic et al., 2011).

TYPES OF NUMERICAL MODELS

Discrete fracture pathways and mass transfer between mobile (generally fractures) and immobile (generally matrix) porosities can be modeled with most hydrogeologic modeling codes. The level of accuracy and effort required for modeling will depend on how well the selected numerical approach can be parameterized and discretized, given financial, technical, and schedule constraints (Selroos et al., 2002). In general, the more explicitly the fractures and rock matrix blocks are represented, the easier it will be to relate the hydrogeologic model to the underlying geology, and less dramatic simplification will be required. Modeling approaches that explicitly represent fractured rock systems include the DFN approach (Long, 1983; Robinson, 1984; Dershowitz, 1985), the hybrid equivalent porous media (EPM)/DFN approach (Neuman, 1987; Bordas, 2005; Dershowitz, 2006; Illman, 2014), and channel network approaches (Watanabe et al., 1997; Bruines, 2003). Reviews of these models and concepts can be found in the literature and include Evans et al. (1987), Haneberg et al. (1999), Faybinshenko et al. (2000), Berkowitz (2002), Selroos et al. (2002), Dershowitz et al. (2004), Ijiri et al. (2009), and Illman (2014). Neuman (2005) provides a broad overview of trends, prospects, and challenges when quantifying flow and transport in the fractured rock environment. Table 4.4 lists numerical modeling approaches for fractured rock. DFN models are available through commercial, government, and academic sources, but are not used as commonly as continuum models even though they offer advantages for modeling fractured rock system.

There is considerable room to improve the ability of even state-of-the-art numerical models to explicitly describe flow and transport in fractured low permeability systems. The Used Fuel Disposition Campaign Disposal Research and Development Roadmap (Used Fuel Disposition Campaign, 2012) highlighted several ways to characterize and model features in a natural system to determine suitability for waste disposal. That document called for improved numerical modeling tools that represent fractures as discrete features, such as DFN and hybrid DFN/EPM multiple-porosity methods. Single-porosity EPM-type models are only applicable for fractured rock systems when the consequences of severe simplification of the system have been addressed. Loosely coupled approaches to examine geochemical, thermal, geomechanical, and biological interactions can be combined with explicit DFN approaches for scoping calculations before resorting to EPM approaches.

Other effective tools are widely available but not widely used, including model-independent parameter estimation and uncertainty analysis (e.g., Sandia National Laboratories, 2010; Austria et al., 2015) and calibration-constrained Monte Carlo analysis of highly parameterized models using subspace techniques (e.g., Tonkin and Doherty, 2009). Given the number of choices, it is critical that the modeler is familiar enough with the range of available approaches to ensure that simplifications made, given project resource constraints and available data, are those that provide the most insight, the lowest uncertainty bounds, and the best possible forecasts, and therefore the most defensible decisions.

Tools for model optimization, parameter estimation, sensitivity analysis, and uncertainty quantification frameworks can be used to improve the predictive power and quantification of uncertainty from all hydrogeologic models. Some of these have been adapted specifically for fractured rock (e.g., Vesselinov et al., 2001; Reed and Hetland, 2002; Finsterle, 2004; Doherty et al., 2010; Vesselinov and Harp, 2012).

TABLE 4.4 Approaches to Numerical Modeling

Numerical Model Approach	Model Description	Sample References
Discrete Fracture Network (DFN)	Fractures are explicitly represented, usually as meshes of 2D elements (polygons) in 3D space or intersecting pipe elements in 2D space to approximate the discrete pathways and heterogeneous connectivity. Heterogeneity on fracture planes can be represented by assigning varying material properties to elements. The rock matrix between fractures may be represented with an EPM approach, using analytical solutions or 1D approximations. Efficiency of this approach arises from the ability to model a 3D volume with 2D elements. This is balanced by computational demands where large numbers of fractures are included and by the complex geometry of fracture intersections.	Long et al., 1982 Dershowitz et al., 1998, 2004 Jackson et al., 2000 Hartley et al., 2006
Equivalent Porous Media (EPM)	Fractured rock represented as a mesh containing cells (elements) with flow and transport properties defined by a continuity flow equation (Bear, 1972). Each element in the mesh connects to its neighbors so flow is continuous. Transport pathways are determined by flow field, not discrete features. The effect of fractures on flow and transport may be represented using anisotropic elements and by varying element properties (e.g., using stochastic or conditioned approaches). When gridded finely enough, variation of material properties can be similar to that of a DFN model and can provide the heterogeneous connectivity seen in fractured rock. EPM model geometry is the same as that of elements used (i.e., 2D EPM models use 2D elements; 3D EPM models use 3D elements). Multiple techniques can be used (e.g., "non-neighbor connections") to approximate connectivity. EPM approaches may discretize both fractures and rock matrix into a single mesh, or may implement separate interacting meshes for fracture and matrix permeabilities (e.g., dual continuum).	Bear, 1972
Channel Network (CN)	A DFN approach in which 1D elements are used to implement the fracture geometry. The rock matrix may be implemented as a CN model or using an EPM approach. This can improve computational efficiency, but can cause problems in converting fracture networks to pipe networks. The approach is attractive where fracture intersections or solution-enhanced permeability are a significant factor in defining flow and transport pathways. CN approaches defined along transport pathways can be computationally efficient for modeling reactive transport (sorption, reactions) and mass transport between rock matrix and channels via idealized channel/matrix interaction models.	Moreno and Neretnieks, 1993 Eker and Akin, 2006
Stochastic Continuum (SC)	Assigns permeability and porosity to the elements of a DFN, EPM, or CN model using, for example, statistical, geostatistical, or geological approaches. Most DFN models and many EPM models incorporate at least some element of SC to account for uncertainty and variability. SC approaches are useful particularly where a single realization of the random fields will not represent a specific site but a well-defined suite of realizations can display a range of behavior representative expected at a specific site.	Neuman, 1987 Blessent et al., 2009
Hybrid Discrete Fracture Network/ Equivalent Porous Media (DFN/EPM)	The hybrid approach generally breaks the model volume into multiple domains in which the DFN approach is used in some volumes and the EPM approach is used in others. These nested models require a numerical strategy to communicate flow and transport between DFN and EPM domains.	Dershowitz, 2006 Bonneau et al., 2013

TABLE 4.4 Continued

Numerical Model Approach	Model Description	Sample References
Dual Continuum/ Discrete Fracture Network/Equivalent Porous Media (DC/DFN/EPM)	This approach is similar to the hybrid DFN/EPM, but a dual continuum approach is used in the continuum portion of the model.	Pruess, 1985 Tatomir et al., 2011

UPSCALING AND MODEL SIMPLIFICATION

The upscaling process simplifies model properties for application in larger-scaled models. Even single-porosity EPM approaches are implicitly upscaled from the actual geology. The properties of grains and pores are simplified to parameters such as "hydraulic conductivity." In fractured rock models, fracture plane properties need to be simplified. There are limitations to upscaling in any application that may affect the result of analyses.

By upscaling some properties, numerical models can focus on other hydraulically important hydrostructural properties and features. When using an EPM, consideration must be given to how the selected parameters relate to the geometric and hydrodynamic properties of the underlying fractures and rock matrix, and how simplification to a single-porosity EPM model might avoid misleading results. At a minimum, the sensitivity of model results could be evaluated within the range of model parameters and boundary conditions consistent with the hydrostructural model.

Powerful, easy-to-use, and simple approaches are available to upscale fracture hydrostructural models and DFN to heterogeneous and anisotropic EPM models. These make it possible to respect the spatial pattern, heterogeneity, connectivity, and fracture/rock matrix interaction of a realistic hydrostructural conceptual model in a more conventional EPM model. Examples include the Oda fabric tensor (Will et al., 2005; Wang et al., 2008; Matthäi and Nick, 2009) and numerical permeameters (Lu and Kwicklis, 2012). These methods allow large-scale, standardized EPM approaches to preserve some of the discrete pathway and mobile-immobile zone mass transfer behavior.

The upscaling process for fractured rock requires definition of a "cut-off" scale between fractures that are considered explicitly and those that are included within the effective rock matrix porosity and permeability. Careful consideration is necessary of the cutoff scales between fractures upscaled to rock mass porosity and anisotropic permeability, fractures represented as a stochastic population, and fractures represented deterministically.

Hydrostructural models for many fractured rock sites include heterogeneously connected fractures and interacting blocks of rock matrix with high storage porosity. The important rock connectivity (discrete pathway) and rock matrix storage features need to be preserved during upscaling. If this is not possible, then supplemental analyses need to account for discrete pathway and dual porosity effects. Upscaling approaches are frequently better at retaining spatial heterogeneity and anisotropic permeability behaviors than at preserving underlying discrete pathways and connectivities. Mobile-immobile zone interactions, discrete flow pathways, and flow barrier features need to be preserved.

The shift of focus from properties of specific fractures to an equivalent continuum hydraulic conductivity, storativity, and porosity is inevitably a process of upscaling. Upscaled continuum approaches generally assume that flow streamlines correspond to transport pathways. This assumption can result in severe miscalculation of groundwater travel time, effective plume dispersion, and long-tail retention. In many cases, discrete pathways need to be considered separately.

This can be achieved in some of those cases by supplementing upscaled models with simplified scoping calculations for discrete pathways.

Another potentially inaccurate but common assumption is that hydraulic tests in fractured rock can directly define hydraulic properties such as hydraulic conductivity and porosity that in turn can be used as grid elements properties in a continuum model. Hydraulic tests apply hydraulic loads to specific discrete fractures and fracture network that happen to intersect the well. Better results are more likely if fracture and rock matrix hydraulic properties are derived from the well test. Properties can then be extrapolated to the population of fractures not intersected by the well, and then used to upscale from the fracture network to continuum grid hydraulic properties.

The process of translating a hydrostructural model to a numerical model often requires gridding. Grid assumptions and the details of upscaling affect the results of upscaled numerical modeling. Hydraulic and transport tests on fractures and matrix blocks can provide information that can be applied with reasonable confidence to other fractures and matrix blocks of the same scale. However, if the numerical model grid resolution is too large to resolve model elements at the scale of the hydraulic tests, then sort of averaging of properties is required to represent those elements numerically. Significant error in flow and transport forecasting can result if limitations imposed by model gridding and upscaling are ignored. Simulation of processes influenced by smaller-scale features (e.g., capillary driven flow, filtering, coupled processes, and reactive surface areas) can also be represented inaccurately or lost. Sensitivity analyses of alternate upscaling and gridding methods can quantify those errors and limitations, allowing decisions about the reasonableness of errors introduced by specific methods. If sensitivity studies are not feasible, as a minimum, then model results need to be used with an explicit caveat that the model error and uncertainty have not been quantified.

NUMERICAL MODEL ANALYSIS

Using the simplest model that matches data and observations for fractured rock systems can lead to incorrect results. Measurements often are not available in the quantities, distributions, or scales needed, and data may not reflect important processes that occur. Formal numerical model analysis applied to fractured rock systems quantitatively relates the hydrostructural conceptualization, the equations used to represent the conceptualization, the algorithms used to solve the equations in a numerical model, the model input parameters, and model outputs. It is important not only that numerical models adequately solve the equations that represent the hydrostructural conceptualization, but also that tools that quantify model uncertainties are applied so that decisions informed by numerical models can be informed by quantified uncertainties. Even when extensive data, observations, and experimental results are available, modeling approaches need to balance the value of more detailed numerical models (e.g., of multiple coupled processes) against the uncertainty reduction.

Numerical Testing of Alternate Conceptual Models

Mathematical methods are often designed for parameters with ranges (minimum, maximum) or distributions (normal, log-normal). When change in model output with respect to changes in input are continuous (i.e., derivative of outputs to inputs are continuous), mathematical tools that take advantage of that continuity can be applied. However, in some situations, alternative conceptual models need to be used that result in discrete, discontinuous changes in model behavior. For example, there may be insufficient data to determine whether the fracture intensity is above or below the percolation threshold. If the percolation threshold is not reached, then the system does not have long-range connectivity of intersecting fractures and flow does not occur. Above the percolation threshold, there is long-range connectivity, and a pressure gradient will drive flow through permeable fractures. The system essentially goes through a phase change with distinctly different responses (no flow versus flow). In another example, subsurface stratigraphy may not be constrained well enough to know whether a fault offsets a highly conductive unit enough to cut flow through that unit.

Scenarios such as these can be conceptualized, and their consistency with data and model results can be tested (e.g., via pump tests with observation wells on both sides of the fault). In the best cases, alternate conceptualizations can be eliminated based on comparison of the data that is sensitive to the alternative numerical model output. In any case, however, even when a conceptu alization is rejected, it is important to document and retain analyses that lead to acceptance or rejection of alternate conceptual models to inform future efforts. It is also important to communicate with sponsors, clients, regulators, and stakeholders how to visualize and represent such scenarios. More information on testing of conceptual models can be found in the literature (e.g., Neuman, 2003; Neuman and Wieranga, 2003).

Model Calibration

A productive approach to develop reasonable model inputs is to identify plausible input ranges based on hydrostructural model conceptualizations, run forward models within an automated parameter estimation framework, identify model output that is most reasonably and quantitatively consistent with observations, and identify the input parameter combinations that yield the most reasonable outputs. However, uninformed use of this approach can lead to incorrect interpretations. Many combinations of input parameters may yield output that is a good fit to observed conditions. A good fit, therefore, does not necessarily ensure the model correctly represents the flow and transport conditions. Data used typically for model calibration include heads, flow rates, and geochemical measurements under natural and pumped conditions. Examples of hydrologic model calibrations include study of potential sites for high-level nuclear waste repositories such as Yucca Mountain (Zyvoloski, 2003) and Äspö (Mazurek et al., 2003).

Sensitivity Analysis

In cases where numerical model output changes continuously with variations in model input values, quantitative measures of model sensitivity can be calculated by what is, essentially, computing the partial derivatives of model output with respect to input. This provides a quantitative measure of sensitivity of output to input. However, complex numerical models with large numbers of degrees of freedom often make it impractical to carry out an analysis of each of the systematic changes made to model input. Doing so would involve a full forward run of the numerical model for each input parameter change—an impractical brute-force approach that is inappropriate for all but the simplest of models. Instead, the sensitivity can be quantified by using sampling methods (e.g., Monte Carlo, screening methods) to statistically estimate the response of the multi-dimensional space of model behavior. This approach samples changes in inputs from a known or defined distribution and characterizes output variability in terms of observed output distributions. Again, in models with many input parameters, sampling the high-dimensional space can be computationally expensive, but methods and tools to do this efficiently are available (e.g., Yeh, 1986).

As described earlier, quantifying model uncertainty can have considerable value. Analysis of initial scoping calculations often reveals ways to simplify numerical models when, for example, some numerical model inputs have little effect on output. Quantitatively analyzing the sensitivity—or in this case, the insensitivity—of a model can be powerful because it can reveal what additional data may or may not constrain a model parameter and decrease model uncertainty. This type of insight is more valuable than that resulting from a single numerical model calculation.

Quantitative sensitivity analysis can help to quantify correlated parameters and inform on ways to simplify models, as well as the decisions related to the types of observations and data needed to constrain the model. Sensitivity analysis is affected by issues of non-uniqueness. For example, many alternative combinations of fracture geometry and hydrodynamic properties may explain equally well the measured responses. Additionally, parameters studied in sensitivity analysis may not be those directly affecting observations. For example, many hydrodynamic responses are a function of hydraulic diffusivity (the ratio of transmissivity to storativity), such that sensitivity analysis of transmissivity alone is a poorly posed problem.

Uncertainty Quantification

Model uncertainty can arise from (a) simplifications necessary to implement a hydrostructural model in a numerical model, (b) limitations of the understanding or implementation of physical-chemical-biological processes in the model, (c) errors in the numerical model implementation, and (d) limitations in the match between measurements and model results. Quantifying these errors and uncertainties ensures appropriate application of numerical models and the conclusions drawn from them.

Just as examining the sensitivity of a model to input parameters can guide subsequent model development, quantifying the uncertainties that arise during the modeling process can guide interpretation of model output. Numerical models and the equations they solve are inevitably imperfect representations of the real world that may be biased or inadequate and result in different categories of uncertainty. Quantitative measures need to be developed to determine the likelihood of remediation success (e.g., that remediation strategies reduce concentrations to below maximum contaminant level; that the mass fraction of a contaminant is accessible given a specific remediation method; that the amount of time for contaminant to reach a compliance boundary is reasonable; and that uncertainty can be reduced through collection of additional, potentially expensive, data).

Structural uncertainties can be large and are often the most difficult to quantify and estimate when modeling fractured rock. Parameter (numerical model input) uncertainty is always present and often can be quantified. Because model inputs are usually based on limited observations, they must be interpolated and extrapolated. Formal (statistical data analysis) or heuristic (expert opinion) methods can be used to place bounds and characterize the possible parameter distribution (normal, log normal). This can be done through statistical analysis of the data if enough data are available. When data are limited, expert opinion with sound, documented explanations for the choices can be applied.

Uncertainty from all sources propagates forward through a numerical model in forward uncertainty quantification. Uncertainties in model outputs can be stated in terms of uncertainty in model inputs. In all but the simplest numerical models, the expense (e.g., time, computational intensity) to evaluate a single forward model makes it prohibitive to sample the full high-dimensional parameter space. Sampling methods (e.g., Monte Carlo, Markov chain Monte Carlo, adaptive sampling) need to be used judiciously to reduce the total number of forward solutions of the numerical model constructed to represent the physical system.

ANALYSIS AND RESOURCES

Numerical models can improve decision processes by forecasting system performance. The processes of data collection, hydrostructural conceptualization, numerical model setup and execution, and model analysis is not a rigid and prescriptive one-pass process. A successful analysis process requires continuous reassessment of data, model assumptions, numerical model validity, and analysis of model output relative to data and observations (see Chapter 7). An application of some of the most advanced methods and practices to siting an underground nuclear waste repository in fractured rock is at the Forsmark site in Sweden. Decades of developing methodologies and algorithms to implement numerical models are described by Hartley and Joyce (2013).

Numerical models and associated software implementations will continue to grow in response to need, but it is often a slow process to bring the latest technologies into mainstream practice. Time and resource availability are often inconsistent with the complex and resource intensive workflows required to implement the latest technologies, and the interdisciplinary nature of the analysis of flow and transport in fractured rock will always make that analysis challenging. Many analysis tools are not widely available or are difficult to use and, as such, may not be accessible to everyone. The use of oversimplified or inappropriate numerical models, however, can lead to inadequate or incorrect results. Hydrogeologists that receive proper training that allows them to develop appropriate conceptual models and then apply inexpensive initial scoping calculations to bound system behavior will be able to make more informed decisions regarding data needs and more complex modeling.

5

Methods for Site Characterization and Monitoring

Advances in fractured rock characterization and monitoring in the past 20 years have improved understanding of fractured rock, particularly as it relates to the geometry and transport properties of rock fractures and to fluid exchange between rock fractures and matrix. Characterizing hydraulic properties of rock fractures is now standard in practice and provides critical information about flow paths and physical processes that affect chemical migration. Equally important are characterization and monitoring of groundwater geochemistry, including spatial and temporal variabilities, and experimental procedures that identify the significance of biogeochemical processes in contaminant fate and transport.

A number of methods for detecting fractures and their characteristics were outlined by the National Research Council (NRC) in 1996 and are still used, including wireline logs, tracer tests, seismic methods, and flowmeters (NRC, 1996). Recent advances include automated, longer-term, and autonomous data collection that allow physical, chemical, and biological processes that control the transport and fate of contaminants to be better quantified. A revolution in sensor technologies and low-cost techniques has occurred in the past two decades. This chapter focuses on techniques for use beneath the vadose zone that have evolved significantly over that time.

Significant difficulties remain in characterizing and monitoring fractured rock systems: models and decision making are not always informed by data, data are collected over a wide range of scales, biological process characterization remains in its infancy, and the collection and analysis of those data remain difficult for many practitioners. Characterization at depths greater than a few hundred meters compounds the difficulties because of the need to account for increases in pressure, temperature, and changes in salinity. The lack of accessibility to greater depths means more reliance on remote or indirect measurement of properties. Many of the same surface and downhole methods used commonly for nearer-surface characterization and monitoring, however, are also applied or planned for deep rock characterization (e.g., Arnold et al., 2012). Research related to the feasibility of geologic sequestration of carbon dioxide necessarily focuses often on issues related to characterization and monitoring at depth (see Box 5.1 for a brief description).

Improvements in practice will occur when parameter estimation methods for fractured rock systems and models are widely adopted. Such methods include joint inversion techniques that make it possible to synthesize all available information in large, complex, multi-disciplinary data sets. This chapter begins with a general discussion on the importance of geomechanical characterization and then describes geometric, hydraulic, geophysical, geochemical, and biological characterization techniques. The focus of these latter sections is on areas that show great promise in the characterization of the fractured rock environment.

GEOMECHANICAL CHARACTERIZATION

Geomechanical characterization of fractures is important to understand the geometry, flow, and transport properties of existing natural fractures, as wells as the coupling of fracture flow and transport properties with rock stress and deformation. Conventional rock geomechanical characterization uses empirical methods (Marinos and Hoek, 2000; Zafirovski et al., 2012) to estimate in

BOX 5.1 Characterization of Sites for Geologic Sequestration of Carbon Dioxide

Geologic carbon sequestration involves injection of supercritical (liquid) carbon dioxide (CO_2) into appropriate geologic formations for permanent storage at depths typically between 1 and 3 kilometers. Because injection generally is intended to occur over periods of years and for storage of large volumes of CO_2, effective sequestration requires extensive characterization of the intended reservoir and its surrounding rock units (Kaldi et al., 2009; NRC, 2013a; Bundschuh and Al-Khoury, 2014). Although various types of rock units are considered for geologic sequestration, saline aquifers are used most often and undergo extensive characterization prior to consideration for eventual CO_2 storage (Casey, 2008; Chadwick et al., 2008). The characterization process includes understanding the regional geologic framework and history of the rock units of interest (including natural tectonic or seismic activity and regional structure) and the nature and characteristics of the reservoir and the seal or cap rock. Properties of interest for both reservoir and seal include permeability, porosity, mineralogy, unit thicknesses, temperature, confining pressure, faults or discontinuities, in situ stresses, rock strength, and the nature and composition of preexisting fluids in the reservoir (Kaldi et al., 2009; Eiken et al., 2011; Morgan and McCoy, 2012). Data and interpretation of this kind of information allow development of structural and stratigraphic subsurface maps and models of the target reservoir, which can then aid in predicting the flow and transport of CO_2 in the reservoir through time (Morgan and McCoy, 2012).

Because CO_2 is injected for storage over many years, pre-injection site characterization has a significant role in long-term monitoring of the movement of the CO_2 plume, maximization of the reservoir storage capacity, and minimization of risk. Characterization of CO_2 geologic storage sites employs many tools also used in the oil and gas and geothermal industries. Surface geophysical measurements important for site characterization include seismic (typically including three-dimensional surveys) and gravity, measurements at the well head and downhole (ambient gas levels, well cores, well logs, pressure testing), and regional geology and geologic history of the region. Geologic, structural, and stratigraphic maps can then be generated and linked to information about reservoir characteristics (e.g., temperature, downhole pressure, rock and preexisting fluid composition, faults and fractures, and in situ stresses) (Eiken et al., 2011; Morgan and McCoy, 2012). Once injection begins, continuous monitoring at the injection site involves the use of a range of tools, which contribute data to build on the pre-injection baseline understanding of the rock reservoir. Microseismic analysis, time-lapse seismic and gravity monitoring, electrical and electromagnetic surveying, satellite-based Interferometric Synthetic Aperture Radar and tilt measurements to observe ground deformation, downhole monitoring of pressures and temperatures, groundwater monitoring using tracers, air quality monitoring to detect leaks, and data gleaned from nearby observation wells each contribute to understanding of the effectiveness and extent of the CO_2 storage and allow for better, controlled management of injection and development of the reservoir (Eiken et al., 2011; Morgan and McCoy, 2012). Additional information about methods for characterization and monitoring at carbon capture and storage sites can be found in a best practices manual developed by the U.S. Department of Energy's National Energy Technology Laboratory (NETL, 2012).

situ rock mass strength and modulus. Because rock fracture geometry and properties are a major component of rock mass behavior, it is at least theoretically possible to utilize these empirical rock characterization techniques to estimate fracture geometry and properties. Even if the fundamental processes that govern fracture generation are known, the geometric characteristics of the void space within individual fractures and the three-dimensional spatial distribution of fractures cannot be mapped over large volumes deterministically. Geomechanical information can indicate primary fracture orientations and the likelihood of certain fractures being more open and transmissive to groundwater.

The oil and gas industry employs geomechanical characterization techniques such as microseismic methods for information about fracture geometry and in situ stress (e.g., Okada, 2003; Grob and van der Baan, 2011); borehole image and wireline methods used to determine in situ stress and rock mass mechanical properties (Zoback, 2010); and three-dimensional seismic methods for determining rock mass modulus and anisotropy (Dradjat et al., 2012). Borehole imaging and wireline methods are currently used for environmental site characterization, but the geophysical methods described could also be useful in the characterization of fractured rock sites to inform the design and construction of engineered facilities and contaminant remediation efforts.

Characterization of lithologic geomechanical properties and the evolution of local and regional stress distributions can provide insight into mechanisms of fracture generation and the probable modes of fracturing. Fracture orientation and size, for example, can be defined by the geomechanics of deposition and orogeny, and by the history of tectonic stress and strains. The characterization techniques provide valuable but frequently qualitative and statistical, rather than quantitative and deterministic, information about geomechanical processes and degree of variability.

Fracture geomechanical characterization is an area in need of further development, particularly in the determination of local in situ effect stress fields and fracture geomechanical and coupled geomechanical/hydrogeologic properties. Research is ongoing in the mining (e.g., Bahrani and Tannant, 2011), petroleum (e.g., Warpinski et al., 2013), and nuclear waste management industries (e.g., Liu et al., 2012).

GEOMETRIC CHARACTERIZATION

Systematic fracture mapping is an essential tool to delineate the geometries of discrete pathways and the nature of both fracture pathways and rock matrix. Systematic fracture mapping of the broad geologic setting through visual geometric characterization is an early step of any site characterization. Not all fractures at the 1-meter scale need to be mapped if such fractures are part of a single major shear structure that controls a kilometer-scale flow process; determining which fractures to map are a function of the process(es) of interest. Decisions regarding the need to map in more detail individual or sets of fractures can be better informed as more data are collected and site conceptual models are refined. See Chapter 4 for discussion on modeling and Chapter 7 for discussion on decision making. Information can be garnered through visual geometric examination of surface outcrops, boreholes, and bore cores. Informative fracture mapping always includes information about characteristics such as fracture roughness, infilling, porosity, damage zones, and mineralization.

Efficient fracture mapping begins with large-scale lineament mapping techniques such as evaluation of available orthophotography, light detection and ranging (LIDAR),[1] conventional satellite imagery (e.g., LANDSAT/SPOT), and high-resolution airborne geophysical methods (e.g., magnetics and electromagnetics [EM] such as very-low frequency EM and frequency-domain EM). These methods, combined with suitable image processing, may make it possible to map fracture traces, shapes, spacings, and orientations (for non-planar surfaces) for entire outcrops and may yield quasi-three-dimensional representations of exposed fracture surfaces. Smaller-scale mapping (5–50 meters) may then provide greater detail of mechanical stratigraphy and fault architecture.

Traditional visual means to characterize fractures include simple direct or inferred measurements of fracture size, shape, orientation, and spacing from fractures exposed in outcrops. The total length of fractures may not be determined directly, but trace lengths (i.e., the length of the intersection of a fracture with the outcrop surface) can sometimes be determined with tools as simple as a measuring tape. Fracture size can then be inferred with suitable bias corrections (e.g., Mauldon, 1998; Zhang and Ding, 2010).

[1] LIDAR is a remote sensing technology that combines light (via laser) and radar, illuminating a target with laser and analyzing reflected light with concepts from radar analysis to make high-resolution maps that can differentiate subtle topographic features.

Fracture shape is difficult to map from an outcrop unless the outcrop is three-dimensional (non-planar), but it may be predicted based on visual mapping, as in the case of some layered sedimentary rocks. The orientation of the fracture planes in space typically is defined through measurement of the strike and dip of a fracture (i.e., using a geologic compass, or sampling circle). The spacing between fracture traces belonging to a set of parallel or sub-parallel fractures can be measured along sampling lines that are usually oriented in the direction of the mean pole of the fracture (trace) set. Fracture aperture, roughness, and deviation of fractures from a given plane (e.g., waviness) can sometimes be measured with a ruler on an outcrop. Sometimes the rock bounding a fracture can be removed to expose the fracture surface. In such cases, the deviation from an ideal plane can be measured with tools such as a laser profilometer.

Current characterization and modeling strategies rely heavily on observations from boreholes, especially when characterizing sites for deep sequestration of wastes. Boreholes and bore cores yield important characterization information including fracture orientation (either dip or inclination relative to the bore axis if the bore core was oriented), spacing, and aperture. However, such observations only describe heterogeneities in the fractured rock environment at the borehole wall and may generate biased data due to the impacts of drilling the borehole into the formation. Care is necessary to differentiate between naturally formed fractures and those formed as a result of boring. Because coring boreholes is more expensive than drilling, site-specific approaches should define borehole and sampling strategies. If not precluded by cost, then it might be useful to consider using borings drilled in different directions and angles to help resolve sampling bias issues. Judicious and strategic use of coring to recover rock samples for physical and chemical analyses and fine-scale investigation of lithology and fractures can inform site hydrostructural models. Porosity and mineralogy can be analyzed from core samples, and mineral precipitates or weathering may indicate groundwater flow in specific fractures.

Borehole geophysical methods, especially wireline borehole logging, have been used for many years to characterize the fracture location, lithology, and fluid flow. Imaging methods (e.g., borehole cameras and video; acoustic and optical televiewer logging) can be used to identify fractures along the borehole wall. Borehole diameter logging using mechanical or acoustic caliper methods is also fundamental to quantifying the aperture of fractures and in the choice of hydraulic testing approaches.

Sampling bias needs to be accounted for whether fracture mapping data are gathered from surface or borehole/bore core observations. Visible fractures, for example, may be oversampled and their importance overstated in a hydrostructural model. Discrete fracture networking simulation techniques (i.e., Mauldon and Mauldon, 1997) could be useful to quantify the errors and uncertainties associated with these biases for all fracture mapping and fracture logging applications.

HYDRAULIC CHARACTERIZATION OF FRACTURED ROCK

Boreholes provide not only opportunities to visually inspect the subsurface, but also an environment in which to analyze ambient and dynamic hydraulic conditions, conduct tests to estimate hydraulic properties, and collect water samples to evaluate groundwater chemistry and microbiology. Aquifer tests conducted in boreholes can provide explicit information about select groundwater pathways, but they cannot provide complete information about all possible groundwater paths. Test results need to be coupled with a strong knowledge of the geologic conditions and probable fracture distributions to extrapolate hydraulic conditions over larger aquifer volumes.

Uncertainties in conceptualizations need to be acknowledged if no direct evidence from hydraulic testing is obtained.

Transient hydraulic test response, for example, is controlled by the ratio of transmissivity to storativity (i.e., hydraulic diffusivity) and by the connectivity and geometry of the responding fracture network. In porous medium hydrogeology, assumptions are made frequently about storativity to obtain values of transmissivity from the diffusivity. This can be a source of significant error and uncertainty when applied to fractured rock. It is therefore useful to supplement transient hydraulic test information with other data, such as correlations between hydraulic aperture and fracture size, to better constrain both transmissivity and storativity. Hydraulic test interpretations need to move beyond the assumption that the value obtained from a single transient test is representative of all fractures. Interpretations need to include consideration of the spatial and stochastic variability of fracture properties.

In addition to the characterization of fracture and rock matrix hydraulic properties, it is also important to characterize features that behave as flow boundaries, recharge/infiltration boundaries, and discharge boundaries. Conventional hydrogeologic techniques for characterization of fractured aquifers generally can be applied to these boundary features, but their discrete natures still need characterization for accurate modeling. In the petroleum industry, model boundary conditions are based generally on stratigraphic contrasts in permeability—for example, sealing faults or shale layers are defined as no-flow boundaries. This can be problematic in fractured rock, where individual discrete features may connect beyond these faults or low permeability shale layers. The petroleum industry addresses this problem through pressure transient analysis (Streltsova, 1988).

The geometric and hydraulic properties of fractures intersecting a borehole can vary over many orders of magnitude and over short intervals with respect to the length of the borehole. Furthermore, the borehole itself acts as a highly permeable feature in the formation that connects multiple fractures. If the characterization goal is to understand rock transmissivity and groundwater characteristics over a significant thickness of an aquifer (e.g., to determine aquifer capacity to supply groundwater through pumping), then a borehole itself does not necessarily conflict with test objectives because contributions of individual fractures are not being tested.

In contrast, at sites where groundwater is contaminated, or if the goal is to determine the suitability or effectiveness of the geologic environment for waste isolation, characterization of individual fractures and the groundwater flow paths through multiple connected fractures is critical. Because monitoring needs to account for the discrete and potentially tortuous nature of groundwater flow in fractured rock, the variability in the hydraulic head over the length of the borehole, the transmissivity of fractures that intersect the borehole, and variations in groundwater biochemistry in the fractures need to be characterized. Hydraulic communication between multiple fractures in long open intervals in boreholes can yield hydraulic head measurements that are averaged over many hydraulically significant fractures. Additionally, hydraulic communication can also result in cross contamination (see Box 5.2). Methods to isolate individual fractures or sections of boreholes to characterize localized hydraulic and geochemical characteristics are discussed in the next section.

BOX 5.2 Hydraulic Communication and Cross Contamination in Open Boreholes

Open boreholes can lead to contaminant migration of previously uncontaminated fractures and affect water geochemistry (see Figure). Sterling et al. (2005) report cross-contamination of trichloroethylene (TCE) between fractures through an open borehole in fractured sandstone after only 3 days. A multilevel monitoring apparatus was installed after 3 days, but previously uncontaminated fractures had elevated contaminant concentrations for up to 1 year of monitoring. Contaminant redistribution and longevity in previously uncontaminated fractures was due to groundwater moving through fractures as well as contaminant diffusion into the rock matrix. Furthermore, pumping in boreholes open to multiple permeable fractures can lead to geochemical mixing from multiple sources in the water withdrawn from the borehole.

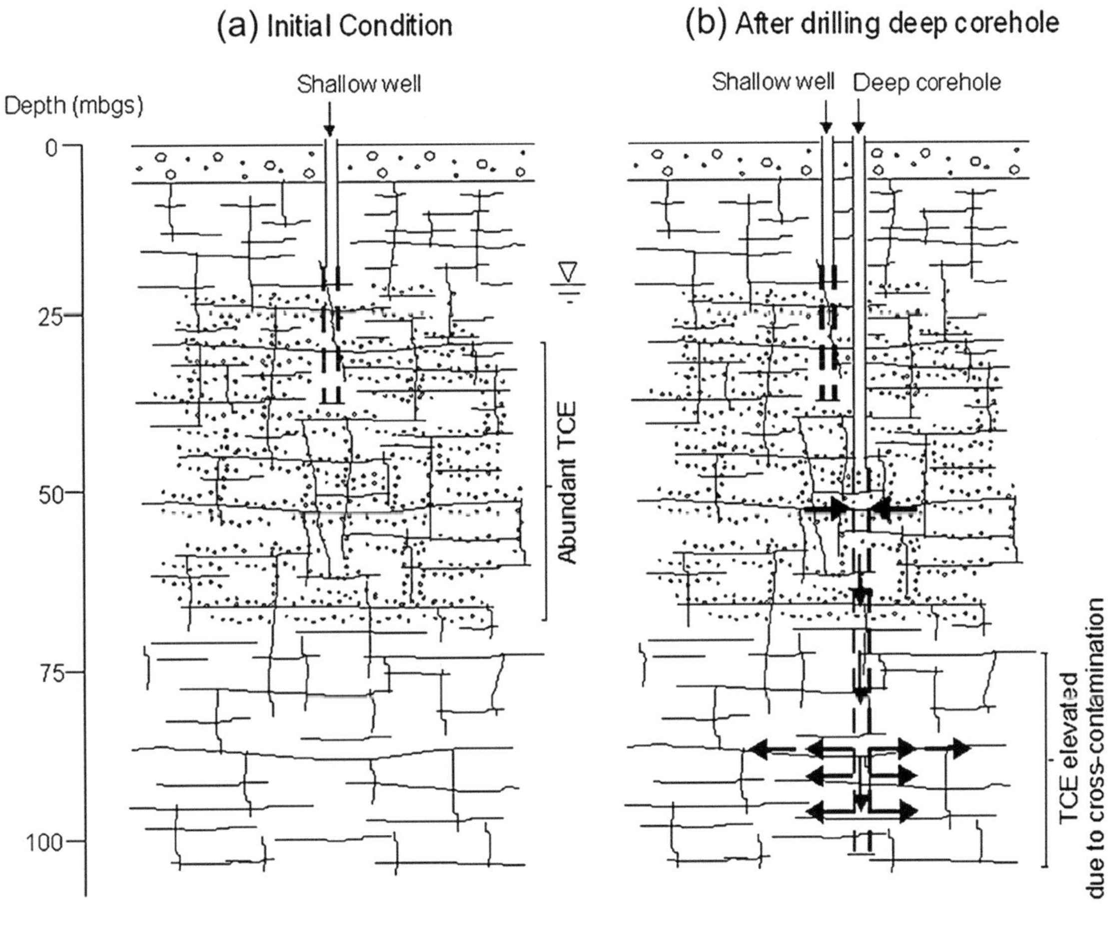

FIGURE Schematic diagram showing the hypothetical distribution and transport of TCE contamination in fractured rock (a) before installation of a borehole and (b) the downward movement of TCE through the borehole and into deeper previously uncontaminated fractures. SOURCE: Sterling et al., 2005.

Some regulatory jurisdictions recognize the potential for cross contamination in monitoring boreholes with open intervals and recommend or mandate the open interval lengths to limit such potential.[2,3,4,5,6,7,8] The lower ranges of recommended interval length are within 5 to 10 feet. A fixed monitoring interval, applied most often when monitoring porous media, is intended to eliminate wider spread of contamination through the monitoring boreholes. Monitoring intervals should be selected based on the synthesis of hydrogeologic and biogeochemical information rather than on arbitrarily assigned values for open interval lengths. They need to be chosen to minimize the number of conductive fractures in a single well screen, avoid connecting flowing fractures of differing hydraulic characteristics (heads), and avoid connecting fractures with different geochemical characteristics (e.g., concentrations, geochemistry).

Isolating Discrete Intervals

Sections of a fractured rock system can be isolated by installing multiple boreholes completed at different elevations and open only over short intervals. Alternatively, pneumatic or mechanical packers (see Figure 5.1) or flexible liners (see Box 5.3) can be installed in a single borehole. When packers are used for hydraulic tests, monitoring of the hydraulic response above and below the packers and of the response in the isolated section is needed to identify connectivity between fractures and to detect leaks around the packers. Perturbing hydraulically (e.g., injecting or pumping water) and monitoring the hydraulic response in the packed-off interval allows estimation of the transmissivity within that interval through a simplified interpretation of the groundwater flow regime (quasi-steady, radial flow) near the borehole. More complex flow regimes associated with single-hole tests have also been developed, but the lack of spatially distributed data related to hydraulic responses along heterogeneous fractures result in great uncertainty in these interpretations.

---

[2] Wisconsin Department of Natural Resources. n.d. *Groundwater Monitoring Well Requirements, Well Screen.* Administrative Code, sec. NR 141.09. Available at https://docs.legis.wisconsin.gov/code/admin_code/nr/100/141/09 (accessed June 19, 2015).

[3] Massachusetts Department of Energy and Environmental Affairs. n.d. *Standard References for Monitoring Wells.* sec. WSC #91-310. Available at http://www.mass.gov/eea/agencies/massdep/cleanup/regulations/wsc91-310-standard-refs-monitoring-well.html (accessed June 19, 2015).

[4] U.S. Environmental Protection Agency, Science and Ecosystem Support Division. 2013. *Design and Installation of Monitoring Wells.* SESDGUID-101-R1. Available at http://www.epa.gov/region4/sesd/fbqstp/Design-and-Installation-of-Monitoring-Wells.pdf (accessed June 19, 2015).

[5] New Jersey Department of Environmental Protection, Division of Water Supply, Bureau of Water Systems and Well Permitting. 2007. *Well Construction and Maintenance; Sealing of Abandoned Wells.* N.J.A.C. 7:9D. Available at http://www.nj.gov/dep/rules/rules/njac7_9d.pdf (accessed June 19, 2015).

[6] Florida Department of Environmental Protection, Bureau of Water Facility Regulation. 2008. *Monitoring Well Design and Construction Guidance Manual.* Available at http://www.dep.state.fl.us/water/groundwater/docs/monitoring-well-manual-formatted-final.pdf (accessed June 19, 2015).

[7] Florida Department of Environmental Protection, Bureau of Petroleum Storage Systems, Petroleum Cleanup Program. 2005. *Design, Installation, and Placement of Monitoring Wells.* Available at http://www.dep.state.fl.us/waste/quick_topics/publications/pss/pcp/MW-SOP-Final-Ap15.pdf (accessed June 19, 2015).

[8] California Department of Environmental Protection, Department of Toxic Substances Control. 2014. *Well Design and Construction for Monitoring Groundwater at Contaminated Sites.* Available at https://www.dtsc.ca.gov/PublicationsForms/upload/Well_Design_Constr_for_Monitoring_GWContam_Sites1.pdf (accessed June 19, 2015).

BOX 5.3 Installation of Flexible Borehole Liners

A flexible liner of the same diameter as the borehole is inserted into the borehole, secured at the top of the surface casing, and filled with water (Figure). When the water level in the liner is greater than the hydraulic head of the fractures intersecting the borehole, the liner is forced against the borehole wall. During installation of flexible borehole liners, water is forced out of the borehole through fractures intersecting the borehole and the liner seals the fractures. The rate of descent of the liner indicates the permeability of the fractures below the bottom of the liner. As the liner seals fractures, the transmissivity of the borehole section below the liner decreases as a function of the liner descent. Transmissivity of individual or closely spaced fractures is estimated by taking the difference between the transmissivity at successive locations as the liner descends the borehole.

Estimating fracture transmissivity is based on the same conceptual model of groundwater flow in the vicinity of the borehole (quasi-steady, radial flow) as is used in the interpretation of flow meter logging and single-hole packer tests. As with flow meter logging in open boreholes, it is difficult to resolve the transmissivity of fractures intersecting the borehole when highly permeable fractures intersect near the bottom of the borehole.

FIGURE Installation of a flexible borehole liner. SOURCE: U.S. Geological Survey; http://water.usgs.gov/ogw/bgas/toxics/FY02-MMR/gallery.html.

Although fracture transmissivities are almost certainly spatially heterogeneous, monitoring perturbations over short durations (minutes to tens of minutes) is intended to represent hydraulic properties of fractures over a relatively small rock volume. Perturbations monitored over longer durations are likely to interrogate heterogeneities within individual fractures as well as those associated with the complex connectivity of multiple fractures in the spatially extensive fracture network. Under those conditions, assuming homogeneity is likely unrealistic. The spatial response to hydraulic perturbation associated with heterogeneous hydraulic properties cannot be captured by monitoring at a single location. A relatively new approach to directly measure the magnitudes and directions of cumulative water and contaminant fluxes without perturbations is use of the fractured rock passive flux meter (FRPFM; Hatfield et al., 2004; Annable et al., 2005), described later in this chapter.

More recently, hydraulic characterization of fractures that intersect boreholes is also conducted using commercially available blank flexible liners (see Figure 5.1c and Box 5.3) to eliminate vertical flow in the borehole. Unlike packers, flexible liners can seal the entirety of the borehole semi-permanently but can be removed if needed. Flexible liners are commercially available but, because of cost, not in general use. Box 5.3 describes how the permeability and transmissivity of fractures can be estimated during liner installation.

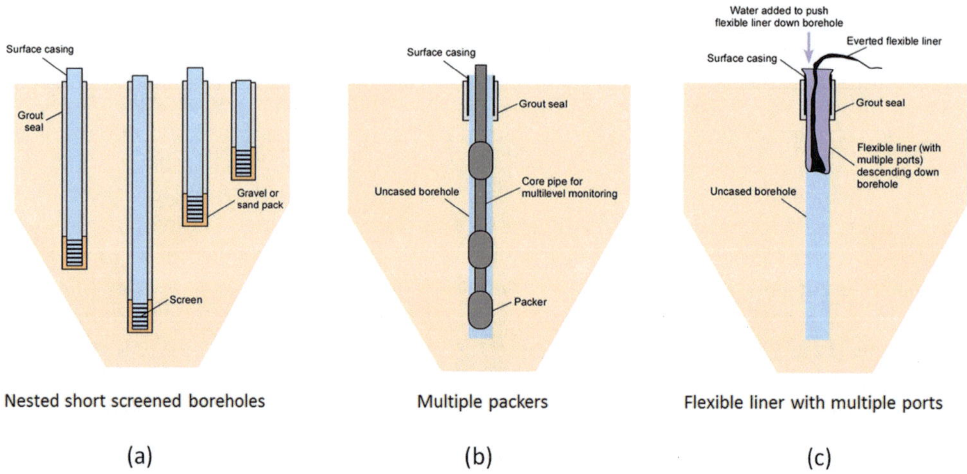

FIGURE 5.1 Means of isolating borehole sections: (a) multiple boreholes open at short intervals and completed at different elevations; (b) multiple pneumatic or mechanical packers in a single borehole; and (c) flexible liners.

Characterization of Multiphase Flow Properties

The characteristic curves that describe multiphase flow in fractures and rock matrix are important for contaminant transport and remediation in both shallow and deep environments. Procedures have been developed to characterize fluid saturations, capillary pressure-fluid saturation, and relative permeability-fluid saturation relationship properties for rock matrix (e.g., Jurgawczynski, 2007), but procedures to characterize these in the fractures themselves are still being developed (e.g., Habana, 2002; Zhang and Fredlund, 2003). The petroleum industry is making significant advances in the use of available data to calculate relative characteristic curves for fractured rock multiphase flow properties (e.g., Lian et al., 2012; Han and Zhao, 2015).

Characterizing Flow Paths at Different Scales

The hydraulic properties of complexly connected fracture systems are sometimes inferred through interpretation of single-hole hydraulic tests using methods that assume fractional flow dimensions (i.e., dimensions of groundwater flow between linear, radial, and three-dimensions). Such assumptions are intended to capture qualitatively the heterogeneous character of the flow regime near the borehole. The fractional flow dimension determined at one location, however, does not necessarily represent behavior at other locations.

Longer-duration hydraulic tests in fractured rock are intended to locate permeable fractures and identify their connectivity over dimensions beyond the immediate borehole vicinity. Hydraulic properties have been estimated over tens of meters or greater using classic aquifer testing approaches since the 1950s and are still used widely for petroleum reservoir characterization. These methods involve hydraulic pumping at a single location under quiescent ambient hydraulic conditions and monitoring the hydraulic responses at a number of monitoring wells. This approach was adopted initially because of computational limitations of interpreting tests from closed-form analytical solutions to groundwater flow equations. Homogeneous aquifer properties or highly idealized heterogeneity typically were assumed. Such approaches have been used widely to characterize fractured rock aquifer properties over sufficiently large physical dimensions that effective hydraulic properties are meaningful. Such classic aquifer test interpretations continue to be used because of their simplicity, but their application should be closely scrutinized when applied to most fractured rock aquifers. When groundwater flow paths need to be defined explicitly to characterize the extent of groundwater contamination, it is inappropriate to rely on a conceptual model with homogeneous aquifer properties or highly idealized interpretations. Additionally, characterizing fractured rock systems at depths of a kilometer or greater offer additional challenges because of their remoteness and because they tend to be subject to larger and more complex stress states that can influence conductive fracture pathways strongly. It is particularly important, then, to characterize the pattern of total and effective stress as spatially varying three-dimensional vectors in these systems (see Box 5.4).

With the advent of numerical algorithms that solve groundwater flow equations in fractured rock systems (e.g., using discrete fracture networks or heterogeneous continua with spatially variable hydraulic properties), hydraulic characterization testing and interpretation can be applied beyond simple and restrictive spatial conceptualizations. For example, at many groundwater sites where contaminated groundwater migration is inhibited through pumping, the spatial connectivity of permeable features needs to be interpreted within the constraints of active pumping because pumping cannot be terminated. However, there may be flexibility and opportunities for gathering information while pumping is under way. Tiedeman et al. (2010) manipulated the pumping rates of a pump-and-treat operation to identify the connectivity of permeable features and estimate their hydraulic properties using available numerical algorithms.

Within a Borehole

Flowmeter logging in bedrock boreholes is a powerful and simple standard practice to quickly interpret the locations of water-producing fractures and to estimate the hydraulic head and transmissivity associated with the most permeable fractures intersecting a borehole. Heat-pulse or electromagnetic flowmeter logging employs a calibrated sensor to measure vertical flow in the borehole under ambient hydraulic conditions and to estimate differences in hydraulic head associated with discrete fractures (usually above and below fractures identified using imaging methods; see Figure 5.2). For example, the heat-pulse flowmeter has been successfully used in fractured rock studies for many years (Paillet, 1998). A packet of hot fluid is introduced into a borehole, and upward or downward movement of the packet is monitored using thermistors[9] to estimate the direction and magnitude of flow.

Recently published algorithms and software enables data collected under ambient and pumping conditions to be combined to estimate the hydraulic head and transmissivity of permeable fractures intersecting the borehole (e.g., Day-Lewis et al., 2011). These interpretive methods assume quasi-steady radial flow in fractures intersecting the borehole. Methods to interpret borehole flowmeter logging data collected under hydraulically stressed conditions have also been developed to infer fracture connectivity between adjacent boreholes.

Although flowmeter logging indicates gross vertical groundwater flow in a fractured rock system, better understanding of relative flows within individual fractures, or fractured regions, the degree of interconnection of these regions, and the spacings of these regions could be gained through semi-quantitative methods. However, unless other approaches are used (e.g., active line source temperature logging, described below), flowmeter logging under ambient and stressed conditions should be employed at all sites during the development of the hydrostructural model.

[9] A resistor, the resistance of which varies greatly with temperature.

BOX 5.4 Fracture Characterization at Depth

The deep subsurface (i.e., kilometer range) is a primary source of both conventional and innovative energy in the United States and a potential storage reservoir for sequestered carbon, energy waste products, and energy resources themselves. Appropriate detailed characterization of the deep surface at a variety of spatial and temporal scales is vital for the safe and sustainable real-time control of subsurface fractures and fluid flow in these systems. Whereas many of the same technologies used in near-surface characterization can be used at depth, their application can be challenging, and new technologies are needed that characterize as they access the subsurface to minimize costs. Innovative drilling, well-construction, and materials (i.e., casings and cements) technologies are needed to make deep subsurface characterization and infrastructure maintenance feasible. New monitoring technologies are required that track the integrity of engineered and natural systems for up to millennia (i.e., for nuclear waste facilities), as are tools for maintenance of the engineered systems at great depths. Durable technologies that allow transmission of data from depth need consideration. The concepts below are of particular importance:

- Larger and more complex stress states can strongly influence conductive fracture pathways at great depths. The patterns of total and effective stress need to be characterized as spatially varying, three-dimensional vectors.
- The occurrence of paleo-fluids at depth, including brines, makes it possible to enhance characterization of fracture storage, connectivity, and transmissivity through evaluation of fracture and matrix pore fluid geochemistry (e.g., total dissolved solids).
- Wireline logs can provide data about the lithology and geometry of fractures at the borehole wall, whereas cross-hole radar and microseismic (where applicable) tomography are particularly well suited to characterizing the geometry of fractures beyond boreholes.
- Borehole image and core data can provide information about the three-dimensional microstructure (i.e., infillings, coatings, and multiple surfaces) that may define important fractures at depth.
- Characterization of porosity and sorption parameters for matrix porosities (i.e., that of fracture infillings, coatings, altered wall rock, and intact rock) that will be accessed over different timescales is important. Time frames of 1 to 100 years are often of interest for fractured rock located near the ground surface. For these time frames, issues such as fracture aperture, fracture coating, and contaminant fate tend to dominate. In deep environments, however, the time frames of interest are often in the range of thousands to millions of years, and diffusion and biological processes become much more important.
- Isotope geochemistry techniques can help with age-dating groundwater to identify residence times and recharge areas.
- Large-scale structures (i.e., a 100-meter-wide fault zone or folds) can be made of multiple smaller-scale structures. Numerous smaller-scale fractures can comprise a fault at any location. It is necessary to characterize the thickness of structures, as well as the fractures and matrix materials within that thickness.
- At depth and at larger scale, the non-planar nature of fractures needs to be understood. When characterizing fractures from boreholes, the orientation of local fracture intersections with wells from the overall fracture orientation need to be determined because they can be significantly different. A 10-degree error can produce a completely different result in terms of potential contaminant migration if it becomes necessary to consider transport at the scales of thousands of meters to 100 kilometers. Of particular concern at great depth is the extrapolation of major structural features between boreholes.
- At depths greater than a kilometer, seismic analyses can identify, for example, changes in rock properties that result from CO_2 displacement of groundwater for carbon sequestration applications. Analyses performed at different times allow monitoring for change (Alshuhail and Lawton, 2007).

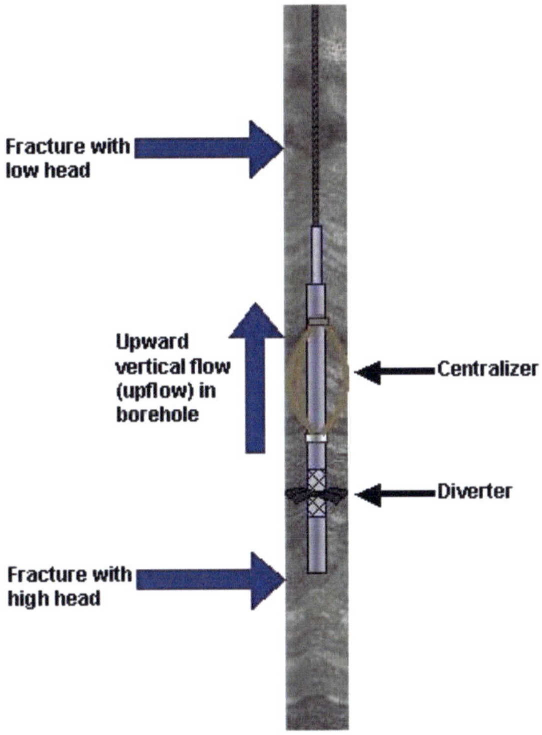

FIGURE 5.2 Schematic of a heat pulse flowmeter in a borehole. The tool is centered in the borehole using a centralizer. Water flows vertically through the borehole from the fracture with higher head to that of lower head and is channeled through the sensors via the diverter. SOURCE: U.S. Geological Survey; http://water.usgs.gov/ogw/bgas/flowmeter.

Meters to Hundreds of Meters

Groundwater flow paths over meters to hundreds of meters can be inferred by monitoring responses to hydraulic perturbations in multiple boreholes. The spatial extent over which a perturbation can be monitored will depend on the strength of the perturbation, the magnitude of the hydraulic diffusivity (defined as the ratio of the transmissivity to the storativity[10]), and the influence of natural groundwater sources or sinks that mask the perturbation. Explicit characterization of groundwater flow paths through hydraulic testing is dependent on the spatial distribution of monitoring wells. However, the cost of borehole installation and completion and, in some cases, the desire to avoid creating additional contaminant migration pathways through the boreholes themselves can discourage additional well installation and, therefore, the ability to identify permeable fracture connections over large volumes of the aquifer.

Regional

Controlled hydraulic perturbations may not be viable at the regional scale, and it may be necessary to rely on ambient hydraulic stresses, regional sources and sinks of groundwater, and a sparse number of monitoring locations. Ambient hydraulic stresses, however, may not be suitable to distinguish preferential flow over regional-scale dimensions because fluid pressure responses from ambient hydraulic stresses are dissipated. The most reasonable option may be to infer bulk

[10] Storativity is defined as the volume of water released from compressive storage per unit decline in the hydraulic head per unit area of the aquifer.

hydraulic properties of the rock from regional groundwater flow modeling. This may be sufficient for characterizing a regional water balance, but it may not accurately portray the groundwater flow regime or address regional-scale chemical transport.

Seismic and Microseismic Characterization

Seismic methods have been used for more than 40 years to characterize large-scale structures (faults) and stratigraphic contacts. Conventional seismic detection methods rely on detection of seismic waves reflected and refracted from a small number of induced seismic events. Characterization of the geometry of such large-scale features (i.e., hundreds of meters) is essential, particularly at depth where these may provide the most significant flow pathways and flow barriers. Recent developments in three-dimensional seismic processing and microseismic monitoring (e.g., Jones et al., 2014) make it possible to characterize the "sub-seismic" scale fractures both deterministically and statistically. Examples of techniques from the petroleum industry include seismic anisotropic processing to characterize fracture intensity and orientation (Treadgold et al., 2008), ant-tracking, and coherence analysis to detect the location and geometry of sub-seismic fractures (20–200 meter scales) (e.g., Pampanelli et al., 2013). These techniques often allow characterization of sub-seismic fractures to significant depths, even in complex geologic settings, and advances enhance the potential of these methods.

Microseismic techniques utilize changes in fluid pressure to detect fractures and have been used in the petroleum industry to detect fractures at resolutions of meters to depths of thousands of meters. These techniques are based on the concept of "critical stress," in which seismic energy is released when fluid pressures bring a portion of a fracture to its shear failure criterion (Zoback, 2007). Microseismic techniques are increasingly used in the mining and petroleum industries from both boreholes and shallow surface installations (e.g., Kilpatrick et al., 2010). Seismic emission tomography (SET) integrates energy over time from many millions of microseismic events (Cornette et al., 2012). This makes it possible for SET to resolve discrete features that are orders of magnitude smaller (meters rather than hundreds of meters), even at great depths and in complex geologic settings. SET is used increasingly in the oil industry, both as an active technique with seismic energy provided by conventional seismic thumpers and hydraulic fractures and as a passive seismic method (PSET), relying on integrated microseismic signals in geomaterials due to ongoing stresses and strains such as earth tides and traffic. The underlying technologies of SET and PSET have been shown to detect fractures to resolutions of 2 to 3 meters at depths of more than 2 kilometers (Shemeta et al., 2012).

Joint inversion of hydraulic and seismic data can be used to improve the characterization of both seismic and sub-seismic fractures. With the availability of more powerful computer resources, the environmental and petroleum industries are advancing this field through the development of new inversion algorithms (i.e., four- and five-dimensional inversions) applied to new types of seismic processing, such as for coherence attributes (e.g., Will et al., 2005; Alcolea and Renard, 2010; Landa and Kumar, 2011; Suman, 2013).

Hydraulic Tomography

Methods to interpret spatially distributed hydraulic properties from multiple hydraulic perturbations—referred to as hydraulic tomography—have been successfully applied in unconsolidated porous media. A variety of algorithms have been used to conceptualize the distribution of hydraulic properties and infer aquifer heterogeneity between monitoring points. The application of hydraulic tomography to fractured rock has been considered (Illman et al., 2008, 2009; Sharmeen et al., 2012; Berg and Illman, 2013; Illman, 2014); however, the resolution of conductivity and storage tomograms depends on the density of pumping and monitoring locations, the quality of data, and how these data are used to produce (usually smooth) maps of conductivity in the inverse problem. To improve the resolution of tomography and therefore the acceptance of these tools, new devices are required that allow higher-density pressure response monitoring at discrete borehole intervals. Additional field trials are needed, particularly those that integrate data from single-hole tests, borehole flow meter profiling, and tracer tests with tomographic results.

Greater Understanding of Flow Paths

Best practices for characterizing groundwater flow paths are described in Box 5.5. Current practice, however, is limited to describing only a small percentage of fractures with the highest transmissivities. Determining the hydraulic significance of less transmissive fracture networks and the rock matrix requires hydraulic monitoring in fractures over a wider transmissivity range and longer duration than currently possible. Groundwater flow likely occurs over a much wider range of transmissivities than hydraulic testing can identify. Expanding hydraulic characterization capacity would clarify the potential for fluid exchange between the most permeable and less transmissive fractures. Over durations of days and months, such groundwater exchange may not be significant. Chemical exchange over decades or centuries, however, may be important.

BOX 5.5 Best Practices for Characterizing Groundwater Flow Paths

Intrusive approaches for characterizing groundwater flow paths (i.e., boreholes) are often limited by the sparse distribution of boreholes. Classical approaches to aquifer testing assume homogeneity and anisotropy and are based on single locations for groundwater withdrawal and injection. They have limited applicability because of the complexity of groundwater flow paths. The best characterization approach, then, is multidisciplinary and integrates tools and the information they provide across individual discrete but often overlapping scales. The best approach also includes

- Designing monitoring apparatus to characterize groundwater flow paths continuously through a borehole via multilevel monitoring. Apparatus designs should be informed by geophysical, hydraulic, and geochemical testing within the boring and should address project objectives.
- Incorporating geologic conceptual understanding (i.e., through the development of a detailed hydrostructural model) into interpretation of hydraulic responses.
- Incorporating complex hydraulic stresses—multiple points of groundwater abstractions (e.g., sites subject to pump-and-treat operation cannot be shut down to accommodate idealized hydraulic stresses).
- Designing multiple hydraulic tests to identify hydraulic connections and preferential flow paths. Algorithms need to be designed that invert hydraulic information and estimate spatial distribution of hydraulic properties consistent with conceptual geologic understanding.
- Characterization of the spatial distribution of flow-limited regions of aquifers important in the retention and release of groundwater contaminants.
- Understanding limitations in ultra-complex geologic settings (e.g., carbonate aquifers subject to dissolution).
- Including uncertainty estimates of parameterizations and overall structure and forecasts.

Geochemical measurements of tritium, radiocarbon, or chlorofluorocarbons that provide information on water age have been used to estimate natural flow paths (e.g., Szabo et al., 1996). Unlike artificial tracer tests, these do not require perturbation of the system. Such analyses have been useful to measure water age and estimate flow paths in porous aquifers (e.g., Szabo et al., 1996; Price et al., 2003). Tritium has been used with great success to define the depth of active groundwater flow in fractured clayey tills (Ruland et al., 1991). Other isotopes have been used in crystalline rocks (e.g., Tweed et al., 2005; Singhal and Gupta, 2010).

GEOPHYSICAL CHARACTERIZATION OF FRACTURED ROCK

State-of-the-art characterization in fractured environments has moved beyond direct observation and imaging of fracture locations and orientations to time-lapse imaging of geomechanical, hydrologic, and biogeochemical processes, leading to a better understanding of heterogeneity, contaminant fluxes, and the contaminant locations. Geophysical methods may allow remote characterization of fractured rock systems at distances from boreholes. This, in turn, may inform more effective and efficient remediation. Geophysical methods, therefore, are worth additional research investment. The methods described in the following sections, beyond those more commonly used and described in NRC (1996), are of particular interest. Many other methods, including seismic, are outlined in detail in that text.

Distributed Temperature Sensing

Temperature data have long been used in hydrologic studies to estimate flow velocities and identify discharge zones. Temperature data are often generated by discrete measurements, either from boreholes or, for example, from stream systems in which baseflow contributions into a stream are controlled by fracture flow. Fiber-optic distributed temperature sensing methods have been used recently and more commonly to monitor temperature in diverse settings. These instruments show tremendous potential to improve practice given their resolution—they are capable of providing temperature measurements over kilometer-scale reaches, with resolution of 1 meter, 1 minute, and 0.1°C (see Box 5.6). Fiber-optic cable is installed in boreholes, streams, or in trenches and attached to a control unit. Although these instruments are more expensive than standard thermistors or thermocouples, they provide spatially and temporally exhaustive data.

BOX 5.6 Distributed Temperature Sensing Systems

Distributed temperature sensing (DTS) systems consist of a control unit and one or more fiber-optic cables. The control unit transmits laser light down the cable and detects the backscatter returning from the cable. Most of the light scatters back at the original frequency, but some scatters at other frequencies dependent on the properties of optical fibers (for an overview on Raman spectroscopy, see Wilson et al., 1980). Light backscattered at a lower frequency is not sensitive to temperature (Stokes band). Conversely, the intensity of light returned at a higher frequency is temperature dependent (the anti-Stokes band). The ratio of the anti-Stokes and Stokes light intensities correlate with the temperature at the specific scatterer along the fiber.

The relative frequency content of the Stokes and anti-Stokes bands can be analyzed using incoherent optical frequency-domain reflectometry (e.g., Farahani and Gogolla, 1999; Park et al., 2006) to yield estimates of continuous temperature along the cable. DTS systems have been used in fractured rock settings to quantify fluid and heat flow through these features (Hurtig et al., 1994; Yamano and Goto, 2005; Pehme et al., 2010; Klepikova et al., 2011; Read et al., 2013). Fiber-optics may also be useful for seismic imaging, thus allowing these cables to be used as continuous geophones (Keul et al., 2005).

Active line source (ALS) logging is another temperature approach to assist characterization via boreholes. In the ALS approach, a borehole is placed into thermal disequilibrium using a heating cable. Temperatures are logged during both heating and cooling using a chain of thermocouples in the borehole. With two or more logs collected during heating or cooling, an estimate of thermal conductivity is obtained. In the absence of groundwater flow in or around the borehole, variations in the thermal conductivity of the rock are due largely to variable water content, and the ALS log provides a reasonable surrogate for a neutron porosity log (Pehme et al., 2007). When groundwater flow dominates the dissipation of thermal energy, the apparent thermal conductivity is increased. In open boreholes this flow can be both ambient (within the formation itself) and connecting (vertical flow between fractures intersected by the borehole). ALS logs are particularly useful to detect ambient groundwater flow in lined holes with no connecting flow. Alternative methods for flow detection, such as chemical dilution or flow meters, require an open borehole and either have poor vertical resolution or require multiple stationary measurements, often with packers to minimize the effects of connecting flow. The ALS technique is a comparatively simple tool—useful in both open and cased or lined boreholes—and run continuously down the length of the borehole. Fracture resolution on the order of a few centimeters.

Nuclear Magnetic Resonance

Nuclear magnetic resonance (NMR) measurements are commonly used in borehole logging to estimate pore size and porosity distributions in porous media. The past decade has brought improvements in sensitivity and quality of measurements that have allowed more routine use in both down-bore tools and core analyzers. These tools likely are suitable for use in fractured rock settings. NMR has been applied at the lab scale to determine fracture aperture (Renshaw et al., 2000) and how fracture wall morphology affects channeling (Brown et al., 1998; Dijk and Berkowitz, 1999; Dijk et al., 1999). Extension to field applications has yet to be accomplished rigorously.

NMR creates a static magnetic field in the borehole that polarizes hydrogen nuclei in water and some contaminants. The magnitude and decay of the polarization of hydrogen nuclei (precession) reveals information about stored fluid volume and type (e.g., gas, oil, or water) and helps differentiate between free and bound water (i.e., attached to rock surface). With careful calibration, NMR may be able to provide information on porosity and permeability (e.g., Legchenko et al., 2002).

The use of NMR is complicated in fractured rock by signal-to-noise ratio. Because the water content is small in fractured rock, the NMR signal is small. The most important control on resolution is the loop size; with very large loops that might be sensitive to up to 100 meters depth, vertical resolution could be expected on the order of meters. The smallest possible loop is probably 25 meters, which might provide sub-meter resolution (Personal communication, A. Parsekian, University of Wyoming, March 29, 2015). In general, the minimum resolvable water content would be on the order of 3 percent, given favorable noise conditions.

Analysis by NMR also has a strong potential to improve our understanding of the behavior of groundwater contaminants in fractured rock. Key to this is the ability of NMR to image the internal structure of core samples such that important features may be identified and mapped. Core analyzers perform a more advanced suite of experiments than well loggers, including capillary pressure analysis, wettability studies, paramagnetic tracer studies, and full three-dimensional magnetic resonance imaging of the interior of cores. The behavior of fluids at these structures may then be observed directly using tracers and other NMR techniques, including the characterization of capillary pressure and preferential pathways for fluid migration. Such studies may improve the models available for the characterization of groundwater behavior at fractures and other lithological interfaces. Further development and testing will be required to determine whether these methods have significant value for testing fractured rock systems and to transition from ex situ, core-based NMR to downhole NMR approaches.

Electrical and Electromagnetic Imaging at the Tens of Meters Scale

Electrical and electromagnetic geophysical methods respond to contrasts in electrical properties across interfaces and have been used to image fractured environments in numerous studies. For example, cross-well, ground-penetrating radar (GPR) tomography (e.g., Olsson et al., 1992; Grégoire and Halleux, 2002), single-well borehole reflection-mode GPR (e.g., Lane et al., 1998), combinations of both reflection and tomography surveys (e.g., Seol et al., 2004), and surface reflection-mode GPR (e.g., Grasmueck, 1996; Tsoflias et al., 2004) have been used to detect fractures. Electrical resistivity (ER) methods have also been used successfully in fractured rock environments to, for example, investigate the dominant fracture-strike direction (e.g., Lane et al., 1995). Resolution of these methods depends on the type and setup of the instrumentation and the geophysical characteristics of the subsurface, among other characteristics; consequently, determining resolution in advance can be difficult.

More recently, these tools have been used to image preferential fluid pathways such as fractures by directly monitoring the subsurface migration of electrically conductive solutes such as sodium chloride tracers. In these cases, changes in electrical conductivity are associated with the movement of a tracer and, when differenced from background values, provide a method to map tracer movement relative to geologic variability and systematic noise. For example, time-lapse, surface-based GPR tomography has been used in a number of studies in fractured rock (e.g., Niva et al., 1988; Talley et al., 2005; Tsoflias and Becker, 2008; Becker and Tsoflias, 2010), although the depth of imaging in these studies is usually limited to a few meters. Day-Lewis et al. (2003, 2004) and Lane et al. (2000) imaged tracer migration at the Mirror Lake, New Hampshire site, although the data in these studies were inverted using a continuum representation of the fracture zones. Lane et al. (1996) and Dorn et al. (2011) have imaged tracer transport using single-hole GPR data. Regardless of whether the authors consider single- or cross-hole measurements, the use of saline tracers coupled with geophysical tools can be useful for meter-scale characterization of transport in fractured rock formations.

ER has been used to monitor transport in fractured rock (Slater et al., 1997a; Nimmer et al., 2007), although less frequently, likely because of the poor resolution of ER when compared to methods such as GPR. Imaging fracture paths in unsaturated materials has also been explored (Slater et al., 1996, 1997b; Zaidman et al., 1999). Recently, new methods have led to improved imaging of fracture features based on alternative mathematics for reconstructing geophysical images that allow for sharp resistivity contrasts at fracture locations (Robinson et al., 2013a,b). These methods show considerable promise for better characterization of fractured rock systems, but they require some knowledge of fracture location from boreholes.

Constitutive Relationships Applied to Field Data

Geophysical data provide indirect measures of rock hydrologic properties. ER, for example, measures electrical conductivity, a material's ability to transmit current, and GPR is largely a measure of dielectric permittivity, a material's ability to store charge. Relationships between geophysical and hydraulic parameters are commonly developed based on the regression of field data (Kelly, 1977; Klimentos and McCann, 1990), empirical studies performed on lab samples (e.g., Topp et al., 1980), and theoretical considerations (e.g., Bruggeman, 1935; Hashin and Shtrikman, 1962; Moysey and Knight, 2004). The derived relationships are assumed to depend on the local properties of the medium and are independent of spatial location. Recent work has shown that a single constitutive relationship may not capture the complexity of a field site given that aquifers are heterogeneous at multiple scales and that the resolution of geophysical images varies spatially (Day-Lewis and Lane, 2004; Moysey and Knight, 2004; Day-Lewis et al., 2005; Moysey et al., 2005; Singha and Gorelick, 2006). Day-Lewis et al. (2005) indicate that geophysical imaging provides "apparent" constitutive relations specific to the geometry, errors, and physics of the survey. Those authors conclude that spatially variable geophysical resolution must be considered to estimate state variables accurately from tomographic images. This issue is rarely considered in practice and may be especially complicated in fractured rock environments.

Geophysical data are not sufficient to quantify hydrologic properties (e.g., transmissivity, tracer concentration and mass) without complementary data to "calibrate" the images produced by tomography (Binley et al., 2002; Yeh and Simunek, 2002; Singha and Gorelick, 2005) or considering joint inversion methods (Kowalsky et al., 2005; Ramirez et al., 2005; Linde et al., 2006; Hinnell et al., 2010). The difference in resolution between hydrologic and geophysical measurements is one of the foremost challenges facing the application of these tools to fractured rock systems. Additionally, characterization and modeling of fluid saturation, capillary pressure-fluid saturation, and relative permeability-fluid saturation relationships are important for geological repository safety assessments in both crystalline and sedimentary rocks. Despite these challenges geophysical methods can be the most effective means to monitor changes in fractured rock systems, if it is understood that they may not provide exact property estimates. A scientifically rigorous and realistic explanation of uncertainty and limitations of parameters derived from geophysical testing is a necessary part of documentation.

Joint Inversion

Joint inversion, or consideration of multiple types of data together to build one consistent model, has led to significant advances in imaging (e.g., Vozoff and Jupp, 1975; Lines et al., 1988; Gallardo and Meju, 2003; Kowalsky et al., 2005). Multiple data sets usually are inverted simultaneously using soft mutual constraints between the data types. Any single technique has significant limitations that may be countered by the different sensitivities of another method. Consequently, integrating multiple data sets may offer the most promise for understanding complicated fractured rock systems.

Although still underutilized for fractured rock systems, joint inversion methods have been used recently, for example, to combine seismic and flow meter data to zone hydrologic models in fractured rock (Chen et al., 2006), ground-penetrating radar data with hydrologic and thermal data to understand coupled processes in a fractured geologic repository for nuclear waste (Kowalsky et al., 2008), and seismic and electrical data to map fracture locations and flow paths (Heincke et al.,

2010). Recent work has also explored the use of flux and head data together in synthetic fracture systems to map fracture distributions and velocity fields in a system with known behavior (Zha et al., 2014).

GEOCHEMICAL CHARACTERIZATION OF FRACTURED ROCK SYSTEMS

Groundwater samples used in the characterization of fractured rock sites are most often collected from boreholes that intersect fractures. Chemical distribution in fractured rock aquifers is often controlled by fracture orientation, transmissivity, spacing, connectivity, and characteristic dimensions of the void space within fractures as well as the void space in the matrix surrounding the fractures. Mobile dissolved contaminants are most often identified with highly transmissive fractures, and their distribution and migration are controlled primarily by the hydraulic gradient and the transmissivity of intersecting fractures. In cases where density is significantly different from background groundwater (high concentrations of dissolved constituents, non-aqueous phase liquids), fracture orientation and connectivity can directly affect migration direction irrespective of hydraulic gradient. When non-aqueous phase liquids (NAPLs) are present in the vicinity of a borehole used for sampling, high contaminant concentrations may be present in lower transmissivity fractures that have been invaded by NAPLs. The presence of the NAPL will reduce the transmissivity until it dissolves, or until NAPL present in higher transmissivity fractures has been depleted through dissolution or diffusion into the surrounding rock matrix. For a review on flow and transport in fractured media, see Bear et al. (2012).

Spatial Distribution of Contaminants in Fractured Rock

Groundwater sampling of individual or multiple fractures provides only a part of the information needed to understand contamination distribution in a fractured rock setting. For example, investigations at the Naval Air Warfare Center research site in New Jersey indicate that less than 1 percent of the total trichloroethylene mass in the subsurface is mobile in groundwater in fractures (see Box 5.7 for a case study). Where contaminants have been in place for decades, a significant amount of contaminant transfer from mobile fluids in rock fractures to the rock matrix can occur. Contaminant concentrations in the rock matrix cannot be ascertained by sampling only mobile fluids in fractures unless the temporal history of concentrations in the fractures is known, and an accurate hydrostructural model at the single-fracture scale has been developed. Concentration isocontours generated from analyzing mobile fluids often cannot be used to quantify contaminants in the subsurface to evaluate, for example, remediation strategies.

Contaminants can be stored in fractured rock matrix and distributed between the rock matrix pore fluids and pore wall rock surface; both can be potential long-term sources of subsequent groundwater contamination (e.g., Grisak and Pickens, 1980). Large concentration gradients between water in fractures and water in rock matrix promote high rates of contaminant exchange between fractures and the matrix. Although less-permeable fractures may not contribute significantly to the volume of groundwater flow, contaminants will migrate from them to more permeable groundwater flow paths as a result of diffusion and slow advection. Similarly, dissolved contaminants diffuse between water in fractures and the rock matrix.

Direct chemical analyses of fractured rock can be critical when assessing the significance of rock surface chemical processes, for example, the organic carbon content of rock can indicate the rock capacity to sorb organic contaminants (e.g., Allen-King et al., 1996). Rock coring, however,

BOX 5.7 Monitoring Contaminant Concentrations in Boreholes Open to Multiple Fractures

The New Jersey Department of Environmental Protection requires monitoring wells at sites with contaminated groundwater to have maximum 20-foot open intervals to avoid extending contamination to other areas of the aquifer. The Naval Air Warfare Center (NAWC), West Trenton, New Jersey, is a trichloroethylene (TCE) contaminated site in fractured sedimentary rocks of the Newark Basin (Lacombe, 2000). The site is characterized by dipping mudstones of the Lockatong formation with groundwater moving primarily through fractures parallel to bedding (Tiedeman et al., 2010).

A particular borehole at the NAWC (36BR) has a grouted surface casing extending 102 feet from land surface and is open for monitoring from 102 to 125 feet below land surface, where there are multiple bedding plane parting fractures. The TCE concentration of water samples collected from the open interval was 83,000 micrograms per liter (Lacombe, 2000). Subsequent investigations used packers to isolate intervals at 102 to 112 feet below land surface (denoted as 36BR-A) and 112 to 125 feet below land surface (denoted as 36BR-B). Single-hole pumping and fluid injection tests were conducted on these intervals to estimate transmissivities. The transmissivity of 36BR-A and 36BR-B was 1.e-5 and 1.e-7 square meters per second, respectively.

Over a 6-month period, TCE concentrations in excess of 15,000 micrograms per liter were measured using a peristaltic pump in 36BR-A. These values were significantly less than the concentration of the open-hole sample (89,000 micrograms per liter). Water samples could not be collected from 36BR-B because low transmissivity in the interval resulted in excessive drawdown incompatible with peristaltic pump use.

The TCE concentration of the open-hole sample was the aggregated effect of two intervals of different transmissivity. The open-hole sample can be interpreted as a flux-average concentration with the contribution of each interval being proportional to the transmissivity of that interval. The open hole concentration can generally be written as

$$C = C_A\left(\frac{T_A}{T_A + T_B}\right) + C_B\left(\frac{T_B}{T_A + T_B}\right)$$

where C is the open hole concentration, C_A and C_B are the concentrations of intervals A and B, respectively, and T_A and T_B are the transmissivities of intervals A and B, respectively. Using the measurements noted in the discussion, the concentration of 36BR-B can be estimated to be in excess of 1,000,000 micrograms per liter, which is greater than the solubility concentration of TCE, indicating that free-phase TCE may exist at the bottom of borehole 36BR. This example illustrates that collecting aggregated (flux-averaged) samples from boreholes in fractured rock open to multiple fractures can lead to erroneous interpretations of water quality in fractures.

is expensive. Relatively few cores are drilled and analyzed, and cores may not provide an accurate sampling of heterogeneities in rock matrix properties or contaminant distribution. It may be decided that rock matrix characterization of this sort is necessary for a site only if there is compelling reason to understand the partitioning between pore water and solid phases. If such a reason exists, then the porosity and organic carbon content of core samples need to be analyzed to differentiate the organic compounds in aqueous and sorbed phases and to develop reasonable rate estimates of sorption and desorption (Sterling, 1999; Sterling et al., 2005; Kennel, 2008). If, however, the additional time to remediate sorbed phase contaminants is inconsequential given the slow rates of diffusion and the resulting decades-to-centuries remediation time frames, then rock sampling may be considered unnecessary.

In the absence of contaminant distribution information, the efficiency and cost-effectiveness of contaminant remediation designs for rock matrix cannot be adequately evaluated. New methods to estimate contaminant mass in the rock matrix using existing boreholes and in situ diffusion experiments are needed. Some geophysical applications may be appropriate for quantifying volumes of less-mobile porosities and rates of exchange (Singha et al., 2007; Day-Lewis and Singha, 2008; Swanson et al., 2012), but additional research in such methods is also needed.

Characterizing Groundwater Chemistry in Fractures

When obtaining water samples from boreholes in fractured rock for chemical analyses, the sample may be dominated by water from fractures that have a limited range in transmissivities (biased to highly transmissive fractures) and readily allow sample collection if the sample interval extends to long sections of the borehole (e.g., Shapiro, 2002). However, fractures not readily sampled (lower transmissivity) also may be contaminated.

Pumping from long open intervals of a borehole results in water being drawn preferentially from the most permeable fractures and yields flux average concentrations. Historical sampling practices have required sufficient pumping to purge the water in the borehole creating an averaged sample across the open interval, with the result that the sample obtained following purging may not represent conditions in any given fracture. On an individual fracture scale, fractures with transmissivity greater than 1e-7 m^2/s will provide sufficient recharge to the borehole to maintain minimal drawdown, but those with smaller transmissivities generally cause reductions in hydraulic head and exceptionally long water sample retrieval times. In a borehole with a long open interval and a mixture of fracture transmissivities, the dominance of the higher transmissivity fractures essentially masks the contributions from the lower transmissivity fractures, which may be the most contaminated (e.g., Johnson et al., 2002). In addition, because many fractured rock aquifers have small fracture porosities, extended pumping to remove standing water in the borehole may draw water from a large volume of the rock through interconnected fractures and therefore the resultant geochemistry may not be representative of ambient conditions in the fractures intersecting the borehole.

As discussed earlier, packer or flexible liner systems can be used to isolate water samples from specific individual or closely spaced fractures, reducing the averaging effect of borehole-scale pumping. Flexible liners can also be used in conjunction with built-in multilevel samplers (e.g., Cherry et al., 2007) that allow the liners to be used to sample water chemistry in individual fractures or in small volumes of interconnected fractures in close proximity to the borehole.

Fractures containing NAPL can also be identified using a flexible liner system formulated to indicate the presence of NAPLs (Griffin and Watson, 2002; Cho et al., 2008), avoiding the need to identify such fractures through secondary lines of evidence (Kueper and Davies, 2009). The liner fabric reacts with the NAPL to produce a stain where it contacts pure product (described in more detail in next section). The use of such liners is also recommended for the "sealing" phase of single borehole investigations where NAPL is suspected to be present.

Several alternative sampling methods have been proposed that do not rely on individual fracture isolation. Sampling from open boreholes by pumping at low rates (low volume; e.g., Puls and Barcelona, 1996), sampling at a given borehole elevation without pumping (no volume; e.g., Savoie and LeBlanc, 2012), and sampling over longer time frames using diffusion techniques (i.e., diffusion-bag approaches—semipermeable bags used to passively collect groundwater samples; Vroblesky and Hyde, 1997) are examples. In some instances, these methods yield similar results to those obtained through discrete fracture isolation; however, discrete fracture isolation is a more robust approach that does not require secondary assessment to confirm that averaging is not occurring to a significant degree. In any of these methods, confirming that geochemical results represent conditions in individual fractures requires that hydraulic conditions in the borehole are understood. If hydraulic conditions give rise to vertical flow through the borehole, for example, then low-volume, no-volume, and diffusion-bag sampling will not be representative of water in an individual fracture, and correlating results to discrete depths is not possible.

An alternate approach without the potential limitations of pumping or diffusion-bag approaches is the FRPFM. The FRPFM is a relatively new technology that measures directly the magnitudes and directions of cumulative water and contaminant fluxes in fractured rock aquifers (e.g., Hatfield et al., 2004; Annable et al., 2005) as well as contaminant fluxes. The design functions under closed-hole conditions across a defined vertical interval in fractured rock wells and is also easily installed in deep rock wells or deep screened wells. The FRPFM directly measures (1) the location of active or flowing fractures; (2) active fracture orientation; (3) direction of groundwater flow in each fracture plane using tracers; (4) cumulative magnitude of groundwater flux in each fracture plane; and (5) cumulative magnitude of contaminant flux in each fracture plane. See Box 5.8 for a description of how the FRPFM functions.

BOX 5.8 Fractured Rock Passive Flux Meter Components

The fractured rock passive flux meter (FRPFM) is composed of a permeable reactive sorbent fabric layer sandwiched between a visible dyed permeable layer and either an inflatable impermeable flexible liner or packer. The dyed layer allows visual identification of the location and direction of flow at the borehole circumference when removed. The figure shows a plan and horizontal cross-sectional view of the FRPFM in a borehole.

The sorbent is a permeable fabric derived from activated carbon, ion exchange resin, etc., depending on the target contaminant. The impermeable liner or packer is made of material typically available in a tube or sock design easily fitted into a borehole of equivalent aperture. The visual indicator layer is a dyed permeable cloth material on the outside of the FRPFM. Once inserted, the FRPFM is inflated, causing it to conform to the shape of the well screen or borehole, with the sorbent layer pressed against the well screen or borehole wall and fractures intersecting the borehole. Hence, the FRPFM is essentially a sampling device with a thin permeable layer of removable sorbent attached to the outside surface of an impermeable flexible liner. The sorptive layer passively intercepts portions of both fracture and matrix flows to simultaneously measure local cumulative solute fluxes and groundwater fluxes. FRPFMs are removed from boreholes days or months after installation, and fluxes are determined from the fraction of tracer lost or the mass of contaminant sorbed (Hatfield et al., 2004; Annable et al., 2005). Because the liner or packer is impermeable, fracture flow does not enter the borehole, but is instead diverted around the impermeable liner. In general, equilibration time of aqueous chemistry is a function of the rock characteristics, well completion, and the drilling method used. Details on timing and protocols for groundwater sampling, broadly, can be found in Barcelona et al. (1985). Passive flux meters are generally recommended to be installed between 2 weeks to 1 month after drilling, depending on the permeability of the rock (Personal communication, M. Tilman, EnviroFlux).

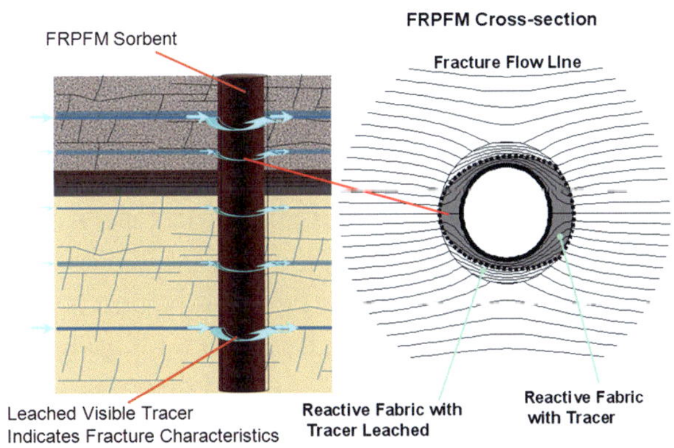

FIGURE Horizontal cross-section of a fractured rock passive flux meter (FRPFM) in an unscreened borehole. The FRPFM is composed of an impermeable flexible liner (e.g., an inflatable or expandable core), a permeable reactive sorbent layer (or fabric), and an outer visual indication layer. SOURCE: Federal Remediation Technology Roundtable; http://www.frtr.gov/pdf/meetings/nov10/presentations/hatfield-presentation.pdf.

The FRPFM for field applications is currently constructed to interrogate a 1-meter interval of the rock borehole. Exposing the FRPFM to flowing groundwater for a selected duration gradually leaches the tracer from the sorbent layer, and the dye from the cloth producing a residual distribution of tracer. Visual inspection of the FRPFM sorbent leads to estimates of location, number, strike and dip, and groundwater flow directions of active or flowing fractures. Further analysis of the sorbent for tracer loss and contaminant accumulation at active fractures will yield the cumulative magnitudes of groundwater and contaminant flux in fractures. These meters have been developed recently for fractured rock flux measurements and applied to a synthetic fractured rock system (Acar et al., 2013).

Characterizing Chemical Transport Processes in Fractured Rock

The spatial distribution of contaminants within a fracture system or at a single borehole can be an indicator of groundwater velocity and dilution only in the rare cases that the original contaminant concentrations and time of subsurface introduction are known with reasonable certainty. Despite this potential limitation, monitoring changes in contaminant spatial distribution may yield qualitative information about groundwater velocity if the processes that attenuate contaminant migration can be accounted for. Groundwater contaminant residence or travel times, dilution, or mixing that occurs during contaminant migration cannot be inferred from single and cross-hole hydraulic tests. Those processes are affected by physical properties of the void space that control groundwater flow, as well as chemical processes such as diffusion and sorption that affect the concentration, spatial distribution, and magnitude of chemical fluxes associated with contaminants in the groundwater flow regime. Characterizing the processes is important in evaluating the fate and transport of groundwater contaminants.

Controlled chemical tracer methods not only are proving successful for characterizing some hydraulic properties and for flow path identification, but also have been successful in the characterization and estimation of chemical transport processes and properties. An extensive outline of the use of tracer tests in fractured rock systems was provided in NRC (1996). In these tests, a tracer is introduced into the groundwater, and arrivals at groundwater withdrawal locations are monitored using multi-level samplers or wells. Given the sparse number of boreholes and the complexity of fracture architecture, tracer experiments should be conducted under hydraulically stressed conditions to ensure that tracers do not simply bypass monitoring location.[11] Under hydraulically stressed conditions, tracer tests can be conducted in single boreholes (e.g., push-pull tests; Haggerty et al., 1998) or between multiple boreholes in close proximity. The results from tests conducted under stress, however, may not be representative of what would be observed in ambient flow conditions. Applying both approaches is preferable but rarely done because of time and cost constraints.

In general, tracer tests conducted under controlled conditions can provide data sets that allow advection and dispersion in fractured rock systems to be estimated. However, large variability in fracture transmissivity along even localized flow paths will lead to tails in the response that are often improperly characterized as controlled by diffusion. Tracer tests are best conducted using multiple tracers in concert to provide resolution of the slow advective and true diffusive effects (e.g., Becker and Shapiro, 2000). At sites where NAPL contamination is thought to exist, partitioning tracers can be employed (e.g., Annable et al., 1998), whereby the tracer partitions into and

[11] Tracer tests have been successful under ambient flow conditions and over great distances when monitoring locations of groundwater discharge in carbonate aquifers (Shapiro, 2011).

out of the non-aqueous phase, resulting in a response at the monitoring point that can be resolved into evidence of the existence of NAPL along the tracer flow path.

In Situ Characterization and Monitoring

Direct, in situ characterization and monitoring of the subsurface is an ideal shared by many sectors interested in subsurface engineering including the oil and gas industries, the geothermal energy production industry, and the environmental management industry. Tracers that provide information about the environment through which they travel may be a promising source of such in situ information for fractured rock. To date, the use of such "smart" tracers for environmental applications has been largely associated with the use of resazurin in surface and ground waters (e.g., Haggerty et al., 2008). Resazurin is a redox indicator to estimate biological activity and has seen groundwater applications such as for characterization for and monitoring of bioremediation (e.g., Guerin et al., 2001).

Various smart tracers and nanosensors are also being used or developed to help in the physical and chemical characterization of petroleum reservoirs (e.g., Anifowose et al., 2013), to monitor geologic storage of carbon dioxide in saline aquifers (e.g., Würdermann et al., 2010), and to characterize in situ stress thermal properties of engineered geothermal systems (e.g., Alaskar et al., 2010, 2011). Use of such technologies in situ, however, will be dependent on factors such as the ability to emplace, protect, locate the sensors, and, in the case of nanotechnology, miniaturize and power the sensors, store data, and retrieve the sensors or transmit data (Matteo et al., 2012). More interdisciplinary research is necessary before these technologies see practical use in fractured rock.

BIOLOGICAL CHARACTERIZATION OF FRACTURED ROCK

Microorganisms inhabit all niches of the subsurface environment, including fractured rocks. The activities of subsurface microbial communities can exert significant effects on both physical and geochemical characteristics and may be responsible for a variety of dynamic processes including mineral formation and dissolution, as well as changes in redox chemistry, fluid surface tension, and pH. Pedersen (1997), Geller et al. (2000), Edwards et al. (2005), and Stoner et al. (2005) used lithographic physical models to demonstrate the significant impacts that subsurface microorganisms can exert on fluid flow in fractures by means of biofilm formation and mineral precipitation.

Methods to characterize microbial communities and activities in fractured rock are similar to those used for porous media, so the discussion here will focus on the most commonly applied techniques with specific examples focused on fractured rock matrices. It is important, however, to note specific challenges associated with proper sampling of microbial communities in fractured rock environments, especially in deep formations (Pedersen, 1997). Drilling into these formations to acquire physical samples can be extremely expensive, and special care is needed to avoid potential contamination of the sampled rock and groundwater from drilling fluids and cuttings. Potential contamination can be overcome by strict adherence to careful aseptic sampling and processing protocols as well as the use of microspheres and tracers to detect drilling fluid intrusion (Griffin et al., 1997; Lehmen et al., 2001). Furthermore, to appropriately analyze microbial activities in environmental samples, they must be maintained under conditions similar to those of their environments, which may include elevated pressures, temperatures, and lack of exposure to oxygen (Geller et al., 2000; Lehmen et al., 2001; Edwards et al., 2005; Purkamo et al., 2013).

Microbial characterization of subsurface matrices is most commonly performed using water pumped from the formation; however, it is important to recognize that many subsurface microorganisms occur as biofilms attached to the solid matrix and therefore may be underrepresented in groundwater samples. Direct characterization of collected solids or down-borehole microcosm methods incorporating beads or other solids (e.g., Bio-Traps) are useful for characterizing the attached microorganisms (e.g., Geller et al., 2000; Lehman et al., 2001; Chang et al., 2005).

Measurements of microbial activity in laboratory experiments generally rely on preserving as many of the environmental conditions as possible, and while they often involve static microcosms, efforts are sometimes made to mimic natural fluid flow, given the importance of this characteristic to microbial activities (Stoner et al., 2005). For example, Geller et al. (2000) applied flow-through "geocosms" created from rock sample cores to study microbial volatile organic degradation in fractured basalt and showed that active bacteria significantly impacted fluid surface tensions and flows. Masciopinto (2007) employed a flow-through system constructed of limestone slabs to evaluate microbial transformation of wastewater nitrogen in fractured formations. Silver and colleagues (2010) constructed flow-through traps emplaced directly within deep boreholes in a mafic sill gold mine to evaluate microbial communities there.

Direct measurements in the field are often used to assess subsurface microbial activities in fractured rock. For example, measurements of oxygen, nitrogen forms, ferric/ferrous iron, sulfur forms, organics, and contaminants can all be useful in assessing microbial activity in fractured rock. In addition, stable isotope ratios of carbon, nitrogen, or sulfur species can provide useful information on microbial activities given that biological processes result in dynamic isotopic shifts in reactants and products. For example, tracking of δ-^{13}C and δ-^{34}S isotopes was used to indicate sulfate-reducing activities within deep granite formations below Aspo Island in Sweden as part of a site characterization for the Swedish nuclear waste disposal program (Pedersen, 1997), and compound-specific carbon isotope signatures were tracked to demonstrate active in situ bioremediation in a variety of fractured rock settings (e.g., Song et al., 2002; Chartrand et al., 2005; Lojkasek-Lima et al., 2012).

Estimates of microbial numbers can be determined for solids or groundwater samples by traditional culturing or molecular techniques; however, it is important to recognize the substantial undercounts associated with culturing techniques of environmental communities (Amann et al., 1995), especially in oligotrophic (nutrient-poor) environments such as those expected within most fractured-rock matrices (Pedersen, 1997). The number of microorganisms detected in groundwater within deep granitic fractured rock environments ranged from 10^3 to 10^7 cells per mL in several studies (Pedersen, 1996; Lehmen et al., 2001), and only a small fraction of these cells were culturable in the lab.

Non-culturing techniques include both non-specific staining methods that target all cells (such as fluorescent DNA stains) and specific staining methods that target individual microbial groups or functions (such as fluorescence in situ hybridization [FISH]). Molecular methods can be designed to be either specific or non-specific such as quantitative polymerase chain reactions (qPCR) that target either universal bacterial or archaeal DNA, or that focus on one specific species, strain, or even functional gene.

The rapid expansion in the past two decades of molecular techniques available to study microbial communities without the need for laboratory culturing has greatly expanded understanding of subsurface microbial communities and the importance they have influencing flow, fate, and transport within fractured rock. For example, Pedersen (1997) highlighted the important role that hydrogen-consuming chemoautotrophs (microorganisms that use hydrogen for energy and inorganic carbon for biosynthesis) likely played in the early evolution of life on Earth and the current role shaping biogeochemical conditions in deep fractured rock.

Molecular techniques involve the analysis of cell-associated DNA/RNA/proteins and metabolites. The most common molecular techniques associated with the study of subsurface environmental microbial communities begin with the extraction of DNA followed by polymerase chain reaction (PCR) to generate sufficient copies of the DNA to be identified and sequenced. Commonly, sequences of genes associated with all forms of life, such as those encoding the ribosomes responsible for protein generation in cells, are used to identify the wide variety of bacteria and archaea present in fractured rock environments. Widely diverse heterotrophic (cells requiring organic carbon/energy sources) and chemolithotrophic (cells using inorganic energy sources) bacteria, for example, were unexpectedly detected in groundwater samples from boreholes in deep granitic aquifers in Sweden (Pedersen, 1997). Molecular techniques were also used to demonstrate the difference in matrix-attached and groundwater-associated microbial populations in an acidic crystalline rock aquifer at Mineral Park Mine in Arizona (Lehman et al., 2001). Purkamo et al. (2013) employed molecular techniques combined with a packer sampling method as a mechanism to evaluate indigenous microbial communities in water drawn from geochemically distinct bedrock fractures in a deep aquifer in Eastern Finland, demonstrating that community structure and geochemistry are strongly linked. Additionally, Lima et al. (2012) used molecular tools to demonstrate that microorganisms present in the rock matrices of a fractured sandstone-dolostone were actively participating in significant dechlorination of halogenated organics.

Recently, a variety of high-throughput molecular techniques have become available to characterize the microbial ecology of complex microbial communities associated with fractured rock. They include nucleic acid sequencing techniques, mass spectrometry based proteomic and metabolomic approaches, and phospholipid fatty acid analysis (Hughes et al., 2000; Venter et al., 2004; Vieites et al., 2009; Bartram et al., 2011; Loman et al., 2012).

The most common high-throughput sequencing techniques include target gene sequencing based on phylogenetic or functional targets and shotgun metagenome sequencing. Both of these techniques investigate the gene content and genetic diversity of microbial communities, but not the functional activity of the cells. For that, metatranscriptomic sequencing of the expressed genes can be performed (Sorek and Cossart, 2010; Moran et al., 2013).

Although recent advances in high-throughput molecular techniques have greatly accelerated the rate of new knowledge development in subsurface microbiology, few comprehensive applications to understand microbial community structure and, more importantly, function in fractured rock environments have been performed. Given the importance of subsurface microbial communities in the geochemical characteristics of fractured rocks, and in the potential for bioremediation of contaminants moving through them, more research in this area is essential and would represent an important investment.

6

Remediation of Fractured Rock

The most problematic contaminated sites are those with persistent contaminants—that is, chlorinated solvents that are recalcitrant to biodegradation in hydrogeologic settings—found in settings characterized by large spatial heterogeneities or the presence of fractures (NRC, 2013b). Furthermore, the deeper below the surface the contamination is, the more problematic remediation tends to be. In the four decades since remediation began in earnest at contaminated groundwater sites, remediation success has been achieved mostly when contamination was shallow, localized, in simple hydrologic settings, and comprised of a single or moderate number of compounds.

In 2004, the U.S. Environmental Protection Agency (EPA) estimated that more than $209 billion would be needed to mitigate these hazards over the next 30 years (EPA, 2004)—likely an underestimate because this number did not include sites where remediation was under way or had transitioned to long-term management. The majority of technologies developed for porous media remediation are not applicable in fractured rock settings (NRC, 1994; SERDP, 2001), and the time frames for remediation can be longer than estimated for unconsolidated porous media sites (without significant contaminant storage) because of slow release of contaminants stored in the rock matrix. Remediation of fractured rock is complicated by, among other things, adequately characterizing the flow system, obtaining representative measurements of spatial and temporal contaminant concentration distributions in groundwater, and the fact that the majority of contaminant may reside in the fractured rock matrix, making the contaminant essentially inaccessible.

Pump-and-treat containment strategies[1] are applied often to remediate many fractured rock sites, but often without detailed knowledge of site hydrogeology, hydraulics, or contaminant location. Although employed frequently, pump and treat is better in the context of containment rather than remediation. Technical details, designs, and monitoring strategies for pump-and-treat remediation are readily found in other works (e.g., Gorelick, 1993; Cohen et al., 1994, 1997; EPA, 1996, 2002, 2008).[2] Over the past decade, in situ techniques such as bioremediation, chemical oxidation, and thermal treatment have been employed increasingly at fractured rock sites. The majority of these approaches (with the exception of thermal treatment) require relatively uniform and predictable contact with the dissolved contaminants, increasing greatly the level of site investigation required as compared to pump-and-treat approaches.

Perhaps the greatest impediment to remediation in fractured rock settings is the lack of common framework, understanding, or expectations related to assessment and realistic remediation endpoints. For example, chemical site investigation data are presented often as contours based on measured concentrations in mobile fluids in fracture porosity (i.e., fractures that transmit fluid). Contours are generated using some algorithm that assumes a linear and structured relationship between data points. Such an approach, however, does not honor the true relationship between discrete data points and is inconsistent with the current understanding of contaminant location if it

[1] Pump-and-treat strategies often involve withdrawal of contaminated water from the ground for treatment. A broader definition of the term was adopted by EPA (1996) to include any remediation system that includes withdrawal of or injection into groundwater.

[2] Information searches on a range of topics are easily performed using EPA's Contaminated Site Clean-Up Information search engine at http://clu-in.org.

is being used to represent remediation performance. A second example is the use of numerical modeling in the design of remedies and prediction of their performance. Historically, little thought has been given to the ability of the underlying equations and assumptions to accurately represent the complex physics of the processes occurring in fractured rock (see Chapter 4). Design underpinned by the use of single porosity models (as an example) is inherently faulty, although that limitation may not be recognized by all appropriate parties.

DIFFICULTIES OF REMEDIATION IN FRACTURED ROCK

Characterization

As discussed in Chapter 5, the level of detail in characterization required for remediation of fractured rock sites exceeds significantly that required at most unconsolidated media sites and may not be possible given budgetary and technology limitations. At the most basic level, any use of source-zone in situ treatment approaches will be challenged by the need to determine the three-dimensional treatment volume. Current hydrostructural models for fractured rock (see Chapter 4) and conceptual models for contaminant distribution indicate the possibility of contaminant penetrations at depths where measurements of depth of penetration of contaminants at fractured rock sites is not trivial. Investigations to hundreds of feet in depth may be required (Parker et al., 2010; see Figure 6.1). Core sample analyses for the entire length of boreholes, in combination with strategically applied non-invasive techniques as described in Chapter 5, may be needed to determine flow distribution and transport and better characterize the depth of the treatment zone. This would be applicable when considering relatively near-surface contamination, as opposed to deep geologic storage or impacts from deep geologic waste disposal. In certain situations characterization that defines treatment volumes may be more costly than the remedy or the point-of-use or point-of-extraction water treatment systems.

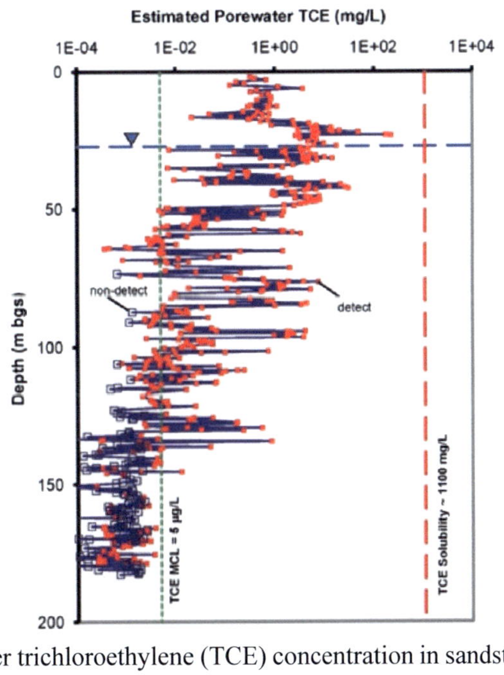

FIGURE 6.1 Estimated pore water trichloroethylene (TCE) concentration in sandstone core samples from California. The concentration is estimated from total volatile organic compound (VOC) mass present in a rock sample (e.g., dissolved, sorbed, and immiscible phase). SOURCE: Parker et al., 2010.

In addition to accurate estimates of the spatial extents of the treatment zone, most remedial approaches that target source zones require an understanding of hydraulics at the individual fracture level for hydraulically important fractures (see Chapter 5). Remediation approaches such as conductive heating can be rendered ineffective if a single highly transmissive fracture is not detected during characterization. Similarly, methods that rely on injection and advection of treatment fluids or amendments will be compromised if a highly transmissive zone is undetected. Rapid migration of injectants and amendments away from the treatment zone can result.

The transmissivity measurement interval used within a borehole can be related directly to estimates of transmissivity. Figure 6.2 shows the measured transmissivity along a borehole through dolomite obtained at different test intervals (Novakowski et al., 2000). Measurements were obtained by packing off different lengths of the borehole and performing the test. Smaller packer spacings may test only a single fracture (or no fractures), whereas larger spacings will average a number of fractures and the interspersed matrix. The figure indicates that measured transmissivity is a function of measurement spacing and demonstrates the extreme variation in hydraulic conductivity that exists in fractured rock settings. This occurs between not only fractures and matrix, but also fractures themselves. This is one of the most critical factors in the design of remediation and ultimately of performance monitoring systems.

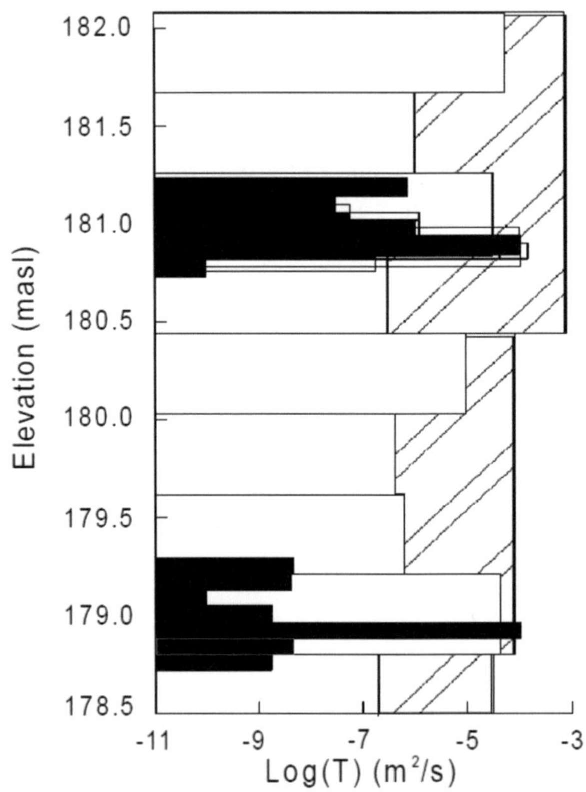

FIGURE 6.2 Variation in transmissivity with scale of measurement in a borehole. The borehole was tested hydraulically with packers at three different spacings: 1.5 m (areas with diagonal lines), 0.4 m (areas with white fill), and 0.1 m (areas with black fill). Measured transmissivity in the top 1.5 m of the borehole ranges from –3.1 (at 1.5 m discretization) to –10.0 (at 0.1 m discretization). Variability in transmissivity within a densely fractured region can be seen between 180.7 and 181.2 masl in elevation where the scale of measurement is close to the scale of the fracture spacing. SOURCE: Novakowski et al., 2000; © Queen's Printer for Ontario, 2000. Reproduced with permission.

Understanding Plumes

The highly variable nature of flow paths in fractured rock settings presents challenges for plume mapping that are rarely of concern in unconsolidated media. In contrast to flow in sands, the migration of contaminants in fractured rock pathways is not resolved easily from coarse monitoring data, and plume migration can dissociate from the averaged direction of groundwater flow interpreted from boreholes with longer screen lengths. Site investigations designed to delineate contaminant plumes, therefore, need to account for the strike and dip of fracture sets when, for example, identifying the location and screen intervals of boreholes. Transport in sedimentary rocks is often along horizontal bedding plane fractures, where "stair-stepping" of contaminant plume is common. A single or series of highly transmissive vertical connecting fractures can result in deviation of some of the plume to higher or lower planes. Vertical gradients and their spatial variations are key controls on the mechanics of plume migration and need to be mapped in conjunction with horizontal gradients.

The depth of contaminant migration in fractured rock (particularly at dense non-aqueous phase liquids [DNAPL] sites) needs to be known to predict the ultimate discharge point of the plume and the extent of groundwater resources that have been impacted. Experience has shown that the majority of plumes in unconsolidated media where the water table is relatively shallow (with the possible exception of sources located within deep groundwater recharge zones) have readily definable discharge points, usually to surface water features such as streams, rivers, or lakes in close proximity to the source of the plume. Investigations of groundwater ages with depth in fractured rock (e.g., Gascoyne, 2004; Palcsu et al., 2007; Takahashi et al., 2013) show that water contained in fractures at depths of greater than 100–150 meters is often on the order of tens of thousands of years old and imply that flow paths from recharge to discharge are potentially basin sized. The process of determining the location, and mapping the flow path, therefore, would require investigations at a regional scale.

The process of diffusion of contaminant into the rock matrix (see Chapter 3) along a flow path results in retardation of the plume migration velocity and a reduction in contaminant concentration. This reduction is often exaggerated as a result of averaging the measured concentrations in samples from monitoring wells with long screen lengths. In such cases, flow in the borehole can be dominated by fractures that are not the most contaminated. The combination of these processes, in conjunction with the difficulty identifying plume centerlines, often results in plumes in fractured bedrock being described as "dilute" or "diffuse." Treatment approaches for plumes in fractured rock conceptualized as large, diffuse, and unstructured are limited if not non-existent.

Access to Contaminants

Remediation at most fractured rock sites is confounded by the large ratio of contaminant in the matrix as opposed to fractures (see Chapter 3). Parker et al. (1994, 1997) demonstrated that the lifespan of a DNAPL in a fracture system can be measured in days to years in many cases, which results in most sites being categorized as late stage by the terminology of Parker (2007; see Figure 6.3). A site in a late-stage configuration has essentially identical contaminant distributions in the source and the plume, and the benefit of contaminant removal from fractures alone is limited.

Remediation approaches such as in situ chemical oxidation (ISCO; e.g., Werner and Helmke, 2003; Goldstein et al., 2004; Shaefer et al., 2012) and enhanced in situ bioremediation (EISB) rely on achieving contact between the oxidant and the contaminant (for ISCO) or the degrading microorganism, the electron donor, and the contaminant (for EISB). Hydraulic injection and advective migration of amendments (such as electron donors or oxidants) will preferentially follow fractures, particularly those of higher permeability. The effectiveness of these approaches is therefore limited because it is difficult to achieve uniform distribution within the fracture system where the contaminants may reside. Furthermore, because much of the contaminant is ultimately stored in the matrix, contact between amendments and contaminants relies on diffusion of the amendment into the matrix—a rate-limiting step in the remediation process.

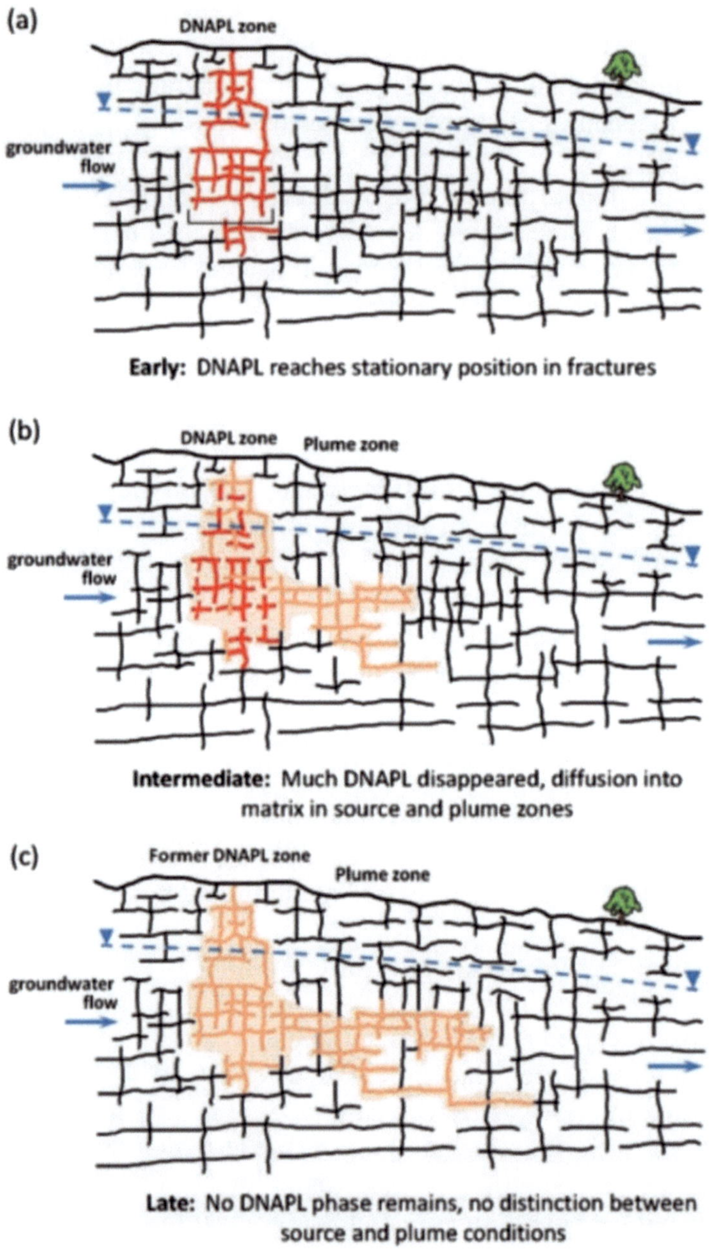

FIGURE 6.3 Life stage of a DNAPL release in fractured rock. SOURCE: Parker, 2007.

Numerous techniques have been proposed for overcoming the rate-limiting diffusion step in oxidation and bioremediation approaches (e.g., use of shear-thinning additives, multiple amendment injections, high concentration amendment injections, use of electrical fields to rapidly migrate amendments into the matrix, hydraulic fracturing, creation of biofilms on fracture surfaces to treat back-diffusing contaminants [see Hill and Sleep, 2002; Charbeneau et al., 2006; Reynolds et al., 2008; Zhong et al., 2011]). Most of these techniques remain unproven in field settings. The use of biological approaches for treatment of contaminant that resides in the matrix may be limited by the ability of degrading bacteria to access, migrate, and survive within the matrix porosity of the rock (see Chapter 3).

POTENTIAL TECHNOLOGIES TO REMEDIATE ORGANIC COMPOUNDS

Kinner et al. concluded in 2005 that the limitations of remediation technologies that are applied to porous media are not known for application in fractured rock. Challenges remain in the form of delivery and distribution of injected material and in remediation of microfractures and low-flow zones in the rock matrix or at very large scales. The same language about the states of knowledge and practice used in 2005 can be applied almost 10 years later. Excluding containment and management remedies such as pump-and-treat and encapsulation techniques (i.e., slurry walls), four remediation approaches remain of potential interest for selected remedial goals at fractured rock sites: bioremediation, ISCO, thermal methods, and monitored natural attenuation.

Biological Approaches

Cost and sustainability indicators make the use of biological remediation in fractured rock attractive. More than 55 percent of sedimentary fractured rock sites contaminated with chlorinated solvents registered by EPA have detectable levels of biodegradation products in groundwater (EPA, 2014a). A significant amount of research and testing has shown that biological treatment is possible within fractures themselves (Hohnstock-Ashe et al., 2001; Lenczewski et al., 2003; Macbeth et al., 2004; Darlington et al., 2008; Bradley et al., 2009; Schaefer et al., 2010). However, limited work has been done on the ability of biological approaches to perform effectively in the matrix porosity of fractured rock.

Historical speculation has been that matrix pore sizes are a limiting factor for microbial growth and transport within sedimentary rock matrices (Lima et al., 2012). However, microbial growth in the matrix porosity of sandstones has been shown in both laboratory and field investigations (Jenneman et al., 1985; Krumholz et al., 1997; Lima et al., 2012). The existence of cells within the matrix porosity of sandstones is not evidence of their ability to contribute to contaminant degradation, and the potential for microbial existence in less porous rocks with smaller pore sizes and less pore interconnectivity is so far unproven. The observed decline in microbial abundance with increasing depth has been associated with various environmental factors; however, the role of geometrical constraints and soil-bacteria mechanical interactions remains poorly analyzed (Rebata-Landa and Santamarina, 2011). Pore sizes may restrict habitable pore space and traversable interconnected porosity (see Chapter 3), and sediment-cell interaction may cause puncture or tensile failure of cell membranes. In addition, because of nutrient and electron donor diffusion limitations, there may be a finite depth to which microorganisms can penetrate into the rock matrix before growth requirements cannot be met (Yu and Pinder, 1994).

Lima et al. (2012) identified three dechlorinators within the matrix porosity of a sandstone contaminated with chlorinated compounds at a distance of 64 centimeters from the nearest fracture. Greater heterogeneity in the microbial population was observed closer to fractures. In general, Lima et al. (2012) found that heterogeneity in microbial populations was present across all samples, even those from closely spaced intervals, indicating the heterogeneity of the rock matrix itself and possibly the heterogeneity of nutrient distribution and availability.

In fractured rock environments, biofilm development has been shown to have a potentially significant impact on network fluid flows (Ross et al., 2001; Hill and Sleep, 2002; Charbeneau et al., 2006; Smith, 2010) through the clogging of transmissive fractures and the resultant diversion of flow into less permeable areas (see Figure 6.4). A biofilm growth layer that is limited to the surface of fractures, and does not clog or significantly reduce the fracture permeability, has the potential to degrade contaminants during the process of back diffusion from the matrix, providing a treatment "barrier" within the individual fractures themselves.

Despite many years of research and field testing, the use of biological remediation approaches in fractured rock environments is still in embryonic stages. Little is known or written about biological remediation for fractured rock at great depth. It is reasonable, however, that biological approaches are possible and should be considered at fractured rock sites. A combination of engineered biostimulation and natural attenuation may provide effective solutions in many cases. The optimal approaches, engineering difficulties, time frames, and realistic endpoints for contaminant bioremediation within fractured rock sites are still poorly understood and represent large research needs.

In Situ Chemical Oxidation and Reduction Approaches

In situ chemical oxidation (ISCO) has a long history of enhancing contaminant removal in unconsolidated porous media. Common oxidants include permanganate, persulfate, Fenton's reagent, and to a lesser extent, ozone. Few Records of Decision[3] include the use of chemical oxidation for source zone remediation in fractured bedrock. Laboratory experimentation on the use of ISCO in fractured settings has shown little impact on dissolved concentrations exiting fractures following flooding; however, contaminant discharge from the fractures decreased, likely because of build-up of oxidation byproducts and resultant decreases in fracture permeability and contaminant transfer rates (Cho et al., 2002; Tunnicliffe and Thomson, 2004; Schaefer et al., 2012). The rate of NAPL removal within fractures during oxidant flooding has been shown to decrease quickly, indicating that NAPL removal is controlled by mass transfer at the NAPL-water interface. However, the overall limits of NAPL removal are controlled by both the mass-transfer rate and the aperture distribution (Schaefer et al., 2012). Research on the interaction of oxidants with matrix-held contamination is extremely limited and has not included physical experimentation. Mundle et al. (2007) found no well-defined correlation between the magnitude of rebound concentrations arising from back diffusion and the percentage of total contaminant destroyed within the fracture. Pang (2010) highlighted that greater than 90 percent of oxidant was consumed by natural organic matter, and the effectiveness of the process was limited by the rate of diffusion of the reaction front into the rock matrix.

[3] A Record of Decision is a public document describing the cleanup strategy to be employed at a Superfund site.

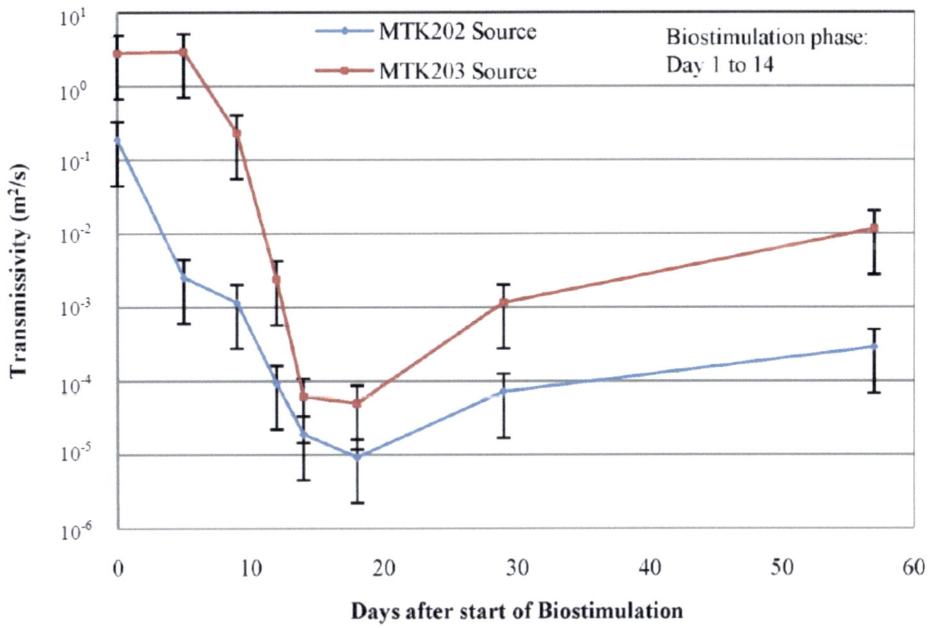

FIGURE 6.4 Fracture transmissivity changes during biostimulation. Transmissivity reduces rapidly with the onset of biostimulation due to biomass growth, and then increases gradually following the termination of biostimulation as biomass decays. SOURCE: Smith, 2010.

The inability of chemical-oxidation approaches to adequately remove contaminant from fractures where DNAPL is present indicates that ISCO approaches are likely best suited to late-stage scenarios (see Figure 6.3). However, the rate-limiting step of the diffusive transport of the oxidant into the matrix porosity will limit the application of the technology, requiring either multiple injections over many years, or delivery methods that can accelerate the migration rate of oxidant into the matrix.

Multiple studies have been published on the use of chemical oxidation in fractured rock at the field scale (Werner and Helmke, 2003; Goldstein et al., 2004, 2007; Helsen et al., 2007). In all cases, difficulties were encountered achieving adequate distribution of the oxidant throughout the treatment zone, rapid rebound of contaminant concentrations to pre-treatment levels, or plugging of fractures and inability to deliver design levels of oxidant.

Gefell et al. (2002, 2003, 2004, 2007, 2008) present the design and preliminary results of a multiple-injection ISCO-remediation pilot project in fractured siltstone and shale bedrock. The intent of the project is to provide permanganate to the fracture porosity through multiple (up to 20) injections occurring approximately every 6 weeks, such that concentrations remain elevated (target in fracture porosity is greater than 20 mg/L) to maximize the potential for diffusive flux into the matrix porosity. Although final results have not been published as of this writing, preliminary results indicate that permanganate transport has occurred up to 100 feet from the injection locations, with peaks appearing from 1–6 weeks following injection. Contaminant concentrations in the target zone and downgradient are decreasing. However, no formal assessment of rebound or degree of penetration and treatment in the rock matrix has been documented.

The use of oxidation and reduction remediation approaches in fractured rock environments is more advanced than biological approaches, but there is still a lack of well-documented case studies, particularly for large-scale implementations. Oxidation and reduction approaches are plausible and need to be considered for remediation of fractured rock sites, but optimal approaches, time frames, and realistic remediation endpoints still need to be determined, and significant effort to obtain permits may be required.

Thermal Approaches

In situ thermal treatment (ISTT) approaches have been developing and applied, primarily in unconsolidated porous media settings, since early trials in the late 1980s. The most commonly implemented versions of ISTT include electrical resistance heating (ERH), thermal conductive heating (TCH), and steam-based heating (Triplett Kingston, 2008). Technical descriptions of those can be found in detail in Johnson et al. (2009). All ISTT approaches involve elevating the temperature of the soil and groundwater within the impacted zone (i.e., source zone or plume) such that volatilization, mobilization, and direct destruction of the chemical compounds occur. Relative to other technologies, some in situ thermal treatment technologies (e.g., ERH) result in preferential heating and contaminant removal from lower permeability media (NRC, 2013b). To date, there are no reported uses of ERH at sites where contaminants are primarily within fractured rock.

The majority (93 percent) of published applications of ISTT since 2000 have been in unconsolidated geologic settings (Triplett Kingston et al., 2010). Steam heating was employed at four of the six fractured rock sites, with TCH and other approaches each employed at one site. Of 14 sites with sufficient data to assess performance in terms of concentration and contaminant flux reduction identified by Triplett Kingston et al. (2010), only one was in fractured rock (described as competent but fractured bedrock), and measured reductions in source zone concentration and contaminant flux immediately downgradient of the source zone were less than a factor of 10.

In comparison to fluid flushing technologies (e.g., oxidant flushing and surfactant flushing), ISTT offers distinct advantages. Heat migration is not as adversely affected by geological heterogeneity as is fluid migration. In comparison to other thermal technologies, TCH has the advantages of (1) not relying on fluid injection (e.g., steam flooding) to deliver heat; (2) being able to achieve temperatures above boiling (impossible with steam flooding or ERH); and (3) being able to destroy contaminants in situ as a result of the high temperatures achieved, thereby reducing the need for ex situ produced fluids treatment. To date, TCH has been implemented in four fractured rock settings:

- A pilot-scale demonstration of TCH in fractured chalk (at the United Kingdom Atomic Energy Authority site in Harwell, United Kingdom) completed in 2005 demonstrated a four-fold increase in soil vapor extraction rate in chalk when the unsaturated zone was heated to ~100°C (CL:AIRE, 2010).
- A pilot-scale test at the National Aeronautics and Space Administration Marshall Space Flight Center Source Area 13 in Huntsville, Alabama, was undertaken in 2007 (Cole et al., 2008). Only the bottom 5 feet of the treatment volume was limestone bedrock, and the majority of the treatment volume was unconsolidated materials.
- At another site in the southeast United States, the total saturated thickness of approximately 25 feet of soil and partially weathered bedrock overlying fractured bedrock was the focus of the treatment zone, with conductive heating occurring approximately 10 feet into the fractured gneiss bedrock (Heron et al., 2008). The effectiveness of TCH on the removal of contaminants from the fractured rock was not demonstrated because of the lack of pre- or post-treatment samples within the rock to quantify contaminant removal.

- TCH in fractured rock was performed at the Naval Air Warfare Center research site in Trenton, New Jersey (Lebrón et al., 2012). Bedrock sample analyses indicate the average reduction in trichloroethylene (TCE) concentrations was 41–69 percent. However, the rock matrix did not achieve targeted temperature in all locations (mostly because of contaminated groundwater influx through existing fractures). Discrete sampling at 5-feet intervals, correlated with observed fractures from borehole video logs, allowed identification of the depth where heating was incomplete and lower levels of contaminant reduction occurred.

Modeling studies indicate careful attention to groundwater influx into a target treatment zone is warranted to determine whether and how long boiling water temperatures can be reached in all locations (Lebrón et al., 2012). Given potential variability of individual fracture flow rates, accurate assessment of the influence of inflowing cold groundwater at the fracture scale is required. Lebrón et al. (2012) indicate that in the case of low groundwater inflow, only quantification of the total groundwater influx through the treatment zone is necessary, rather than characterization of discrete fractures.

Low matrix permeability, high matrix porosity, and wide fracture spacing can contribute to boiling point elevation in the rock matrix. Consequently, knowledge of these properties is important for the estimation of treatment times. This is particularly relevant in low matrix permeability rock where thermal expansion of groundwater leads to pressure increases, which in turn result in elevated water boiling points. Because of the importance of fracture spacing in determining the pressure rise in the matrix, a discrete fracture model is more appropriate than an equivalent porous medium model for simulating boiling.

Predicting Remediation Performance in the Context of Feasibility Studies

Under the Comprehensive Environmental Response, Compensation, and Liability Act, a critical step in the investigation and remediation of contaminated sites is the remedial investigation/feasibility study (RI/FS), which is designed to develop a site-specific, scientifically sound, rigorously supported site remediation strategy. The investigation stage of an RI/FS at a fractured rock site is complex and difficult, given the challenges and costs of data acquisition, data integration, and conceptual model development. The feasibility study stage is additionally challenged given the relatively small number of remediation attempts in fractured rock and the lack of high-quality refereed studies that provide unbiased, critical assessment of remediation success and failure that help to quantify uncertainties associated with various remediation strategies.

The RI portion of the process involves characterization of the site and contaminant, evaluation of risk to human health and the environment, and treatability testing to evaluate the potential performance and cost of the treatment technologies under consideration. The use of treatability studies is well established and understood for unconsolidated sites, but it is less mature for contaminated sites in fractured rock. Properly implemented treatability studies provide key data for the detailed consideration of remedial alternatives, as well as indications of the potential limitation of the remedial alternative being tested. Common treatability studies include reactor testing for bioremediation rate determination (Shah et al., 2009), column flow-through studies for assessment of interactions with site soil or rock (Schaefer et al., 2011; Simmons and Neymark, 2012), batch studies for chemical oxidant effectiveness and consumption (Helsen et al., 2007), and thermal properties testing (Bastona and Kueper, 2009).

Treatability studies can provide the data required to optimize process engineering, determine likely duration of treatment, and forward the design to considerations of scaling at pilot and full scales. Implementation of treatability studies for fractured rock sites requires careful consideration of the differences between laboratory and field conditions. Treatability studies using crushed rock or artificial representations of the rock (induced fracturing) will not replicate the fracture surface area to rock ratio in situ, nor will any flow field established in a laboratory setting accurately reflect the conditions in the field. Treatability studies can and should form a part of any remedial alternatives investigation at fractured rock sites. However, they must be carefully planned and their limitations fully understood and documented.

The FS portion of the process is the mechanism for the development, screening, and detailed evaluation of alternative remedial actions. Pilot testing is sometimes undertaken in the FS stage, particularly if the technology is new or being implemented in a new environment or on a new contaminant. Although not documented, experience on the part of committee members indicates that pilot studies in fractured rock are often costly, difficult to design, and prone to less success than in unconsolidated porous media. Even well-designed pilot tests following significant site characterization and treatability testing (Goldstein et al., 2004; Pearson et al., 2004; Lebron et al., 2012) can fail to meet expectations.

Very little is being published in the peer-reviewed or publicly accessible literature on pilot tests at fractured rock sites. The lack of accessible material does not necessarily indicate that pilot studies are not being undertaken, but it might be taken to indicate a lack of success and, therefore, little drive to publish. In a review of 393 published case studies (FRTR, 2007), none was identified as fractured rock sites. Pilot studies at contaminated fractured rock sites are of value and need to be performed in a rigorous design and execution framework, subject to independent peer review and published to avoid repetition of previous mistakes by practitioners.

POTENTIALLY APPLICABLE APPROACHES FOR RADIONUCLIDES

Radionuclides are naturally decaying inorganic elements that migrate in the subsurface similarly to inorganic materials such as salts and metals. Positively charged radionuclides tend to react with and be sorbed by mineral constituents of rock and, therefore, migrate more slowly than flowing groundwater. Some radionuclides, such as tritium, are nonreactive and migrate by advective transport at essentially the same rate as flowing groundwater. Some radionuclides may sorb to colloidal particles and undergo transport relatively unaffected by geochemical processes. Still other radionuclides are disposed of in the subsurface with high or low pH and may undergo a myriad of reactions that result in precipitation or dissolution. Thus, essentially the full range of geochemical phenomena that one might expect for inorganic contaminants applies to radionuclides. The fact that certain elements might be radioactive affects their potential to affect human health, but in and of itself has little effect on mobility in the subsurface. Treatment of radionuclides removed from the subsurface during remediation may require special technologies and requirements for ultimate disposal.

NATURAL ATTENUATION

Natural attenuation is the reduction in concentration of contaminant in groundwater as a result of naturally occurring biological or chemical processes, dilution, or evaporation that may occur under the correct environmental conditions. Monitored natural attenuation (MNA) is a strategy often applied in the cleanup of chlorinated solvents, petroleum hydrocarbons, and other hazardous wastes in conjunction with source zone cleanup. Biodegradation is an important form of natural attenuation, relying on naturally occurring microorganisms to cause or result in the breakdown of various contaminants into less harmful or inert components.

Matrix diffusion can be considered a natural attenuation process, because it attenuates the rate at which contaminants migrate in the forward direction, and it attenuates the contaminant discharge into the mobile fluid in the reverse direction. Dispersion is also a significant process occurring within fractured rock settings, particularly in crystalline rocks containing multiple hydraulically connected fracture sets resulting in the dispersion of a given contaminant through an ever increasing volume along the flow path.

Natural attenuation through matrix diffusion and dispersion may be less strong in weathered sedimentary systems than in crystalline systems, particularly where fractures are laterally extensive and have large apertures, yet matrix porosities remain low. Plumes generated from strong, high-contaminant sources may grow to significant lengths under such circumstances and only terminate at natural discharge points such as surface water bodies, or at artificial discharge points such as extraction wells. An example of this situation is the Cayuga Springs Superfund site in New York State, where a high strength TCE source exists in a highly transmissive limestone unit approximately 150 feet below the ground surface (EPA, 2012). The resultant plume from this source extends more than 5 miles to a discharge point at a surface water body. Both biotic and abiotic transformation of TCE to daughter products is occurring, primarily because of the high levels of natural organic carbon in the limestone unit. However, the levels of natural attenuation are not sufficient to reduce the contaminant discharge at the receptor to a point where there is no unacceptable risk to human health.

IMPORTANT CONSIDERATIONS IN FRACTURED ROCK REMEDIATION

A conundrum in setting remedial goals, criteria, or even objectives for fractured rock remediation is the potential that the most basic underlying assessment metric—point concentrations in mobile fluids in fracture porosity—is entirely inappropriate. Proper characterization of risk, long-term liability, and performance may require the use of alternate metrics, or additional metrics in conjunction with point source concentration measurements.

Meeting drinking water standards at the points of monitoring, or at actual or potential receptors, is the long-term goal of remediation at most hazardous waste sites in the United States. However, established drinking water maximum contaminant levels (MCLs) are not likely to be met for decades or perhaps centuries for contaminated sites in fractured rock settings. Because U.S. regulatory bodies are becoming more accepting of such long time frames, it is also becoming practical to include alternate approaches to setting goals and objectives for remediation, particularly at fractured rock sites (EPA, 2011). EPA has issued draft guidance that attempts to address setting remedial objectives and formalizes the transition from active remediation to long-term monitoring (EPA, 2014b). The National Research Council (NRC) also discussed this issue, referring to it as a "transition assessment" (NRC, 2013b).

Underlying the difficulties of setting scientifically based, reasonable, and appropriate objectives is the lack of reliable, open-access, peer-reviewed data from previous experience to support the process of constructing objectives. There have been few technical demonstrations of remediation approaches—and particularly as they vary between different settings—to provide a baseline for understanding the performance of the various technologies. Similarly, there have been insufficient controls available to assess the actual long-term performance of remediation efforts in fractured rock, particularly as the performance assessment relates to incremental change in secondary performance indicators (such as contaminant transfer between porosities downgradient of the treatment volume). The confluence of these limitations results in conclusions that are often based on inference as opposed to adequate and appropriate quantitative measures of performance.

This chapter presents a series of key concepts important at an overview level (as opposed to a detailed technical level). Multiple reference sources exist for detailed engineering design, scientific theory, and policy considerations, and the reader is directed to these other more appropriate sources for detailed considerations.

Concept 1. Remediation focused solely on removal or destruction of contaminant in fractures is futile.

As has been discussed in this report, the majority of contamination is often found in the matrix of fractured rock, and this contamination will not be remediated with traditional approaches that focus on remediation of groundwater in rock fractures. Matrix diffusion will dominate the source term in most contaminated settings and will result in plume lifespans that could reach thousands of years. To reduce contaminant concentrations at a given location, remedial approaches must focus on the contaminant held in the matrix porosity of the fractured rock (Parker et al., 2010).

Concept 2. Contaminant source zone remediation has little or no impact on plume lifespans in settings with high matrix porosity.

A remediation effort focused solely on the contaminant source zone in a fractured rock setting is not likely to have a discernible impact on concentrations in downgradient plumes for tens to hundreds of years in rock with high matrix porosity. Parker et al. (2010) used numerical modeling to demonstrate the impacts of source removal on downgradient plumes (see Figure 6.5). A comparison of panels (c) and (d) in Figure 6.5, where (c) represents contaminant concentrations 50 years after complete source removal and (d) represents contaminant concentrations at the same point in time without source removal, demonstrates the calculated marginal impact on the plume. A similar conclusion was reached by Lipson et al. (2005), the main point of which is shown in Figure 3.6 in Chapter 3. Plume lifespans are shown to exceed 1,000 years following complete source removal.

Concept 3. Significant benefits in terms of plume lifespan reduction are more likely through treatment of contaminant within the fractured rock matrix.

Current conceptual models of contaminant distribution in fractured rock settings indicate the importance of contaminant remediation in both the matrix and fracture porosities. This extends beyond source zone remediation to include plume remediation and is a significant difference from remediation in porous media. To achieve significant reductions in concentrations within the mobile fluid in fractured rock at distances from the initial contaminant source (i.e., at a receptor down-gradient from the point of contaminant release), contaminant removal from the fracture porosity along the flow path from the point of release to the receptor is necessary.

The requirement for treatment of fracture porosity along flow paths has a direct and significant correlation with complexity of remedial design, cost, and resistance from responsible parties. In the majority of cases, given the resources required for such extensive remediation and the technical difficulties of even achieving such remediation, a careful and considered assessment of the ultimate remedial goals is necessary.

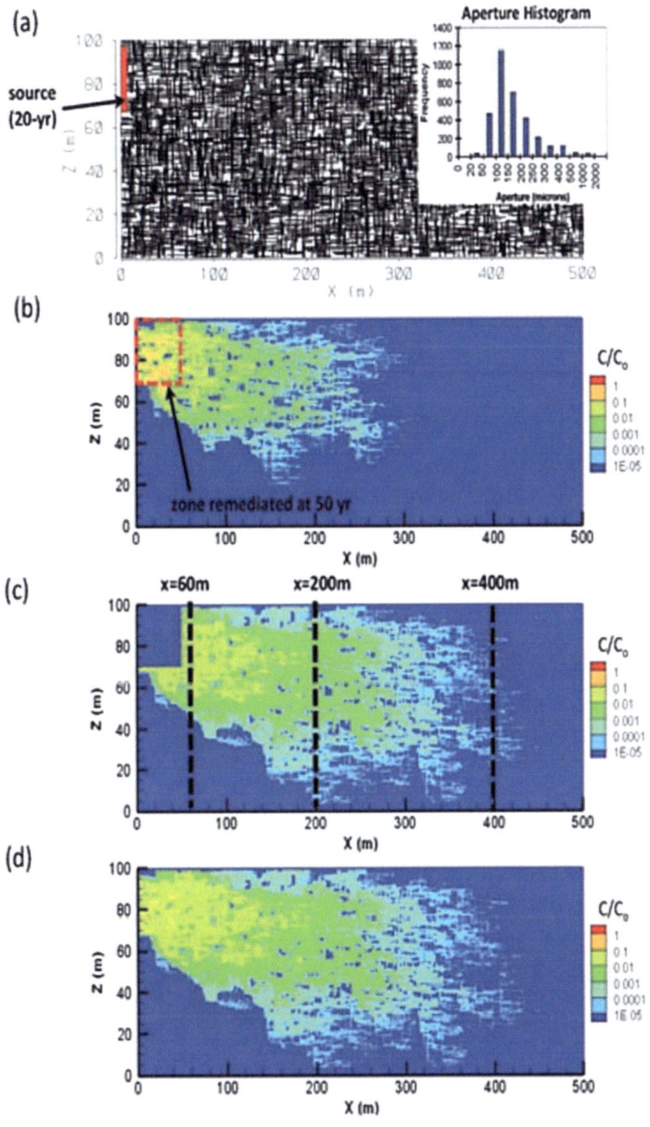

FIGURE 6.5 Results of a numerical simulation of plumes with and without source removal. (a) The model domain, fracture network, and aperture histogram with a constant source for 20 years. The simulation of the calculated plume at (b) 50 years (at which time the source zone is removed), (c) after an additional 50 years following the removal of the source zone, and (d) at the same time as (c) with no source removal. SOURCE: Modified from Parker et al., 2010.

Concept 4. Better characterization leads to better remediation design, better monitoring, and, ultimately, more successful remediation.

The difficulties inherent in fractured rock characterization (see Chapter 5) and the increased cost as compared to characterization in unconsolidated media combine to increase the potential for poor characterization in all stages of investigation at fractured rock sites, especially at large depth. It is well recognized that conceptual site model development needs to be a dynamic process, with new information feeding back to the existing conceptual model and revised conceptual models informing additional investigations (see Chapter 7). This process is critically important in fractured rock settings, and failure to embrace the process is likely to lead to poor remediation design and failure to meet remediation objectives.

A dynamic conceptualization process should be the default practice among practitioners at fractured rock sites (see Chapter 7). However, evidence indicates that either the practice is not followed or the understanding of the process endpoint in terms of uncertainty reduction and provision of adequate, representative, and fit-for-purpose data is limited. An overall indication of the lack of adequate characterization (in both fractured and unconsolidated media) is provided by Triplett Kingston et al. (2010), where only 14 of 182 thermal remediation projects had adequate infrastructure to assess success, indicating that initial conditions for the rest of the projects were not resolved with sufficient certainty to provide a baseline.

The potential impacts of very small-scale site features in fractured rock settings increase the importance of adequate characterization at relevant scales. A single fracture in an anomalous location, or with an "outlier" transmissivity not readily apparent from core examination, has the ability to severely degrade the performance of a remediation system (such an occurrence was documented during a thermal pilot test; Lebrón et al., 2012). A dynamic conceptualization process enhances understanding of where remediation efforts could fail as a result of incomplete characterization, allows examination of failure mechanisms unique to an individual remedy, and can inform a more cost-effective detailed site investigation that, in turn, leads to more effective remedy design.

Concept 5. Better characterization and performance monitoring leads to more honest appraisal of success and failure.

Sufficient characterization of the treatment zone and the larger likely monitoring zone surrounding it is critical for realistic assessment of remediation success and allows for realistic discussion with owners, operators, regulators, and other stakeholders. Degradation of remediation performance that results from a single feature, as described above, is a logical performance monitoring point during and following remediation. Failure to monitor such a feature, which may transmit a high percentage of contaminated flow through the treated volume, would bias performance monitoring and would provide non-representative data for remediation performance assessment.

Concept 6. Natural attenuation is a required component of all remediation solutions in fractured rock.

Because of the skewed distribution of stored contaminant toward the matrix porosity, the difficulty of treating contaminant within the matrix porosity, and the difficulties inherent in the identification and location of fluid flow within the fracture porosity, natural attenuation is likely to be a required component of any remedial approach undertaken in fractured rock. Natural attenuation processes occur in fractured rock to greater and lesser degrees than in porous media, and an understanding of the effectiveness, magnitude, and implications of these processes are crucial to the proper incorporation of natural attenuation into remedial approaches. Effective characterization and parameterization, consideration, and explicit inclusion of matrix diffusion and dispersion as a component of remedial approaches will increase the potential for achieving remedial goals and may even play a role in the setting of realistic and achievable remedial goals at fractured rock sites.

Concept 7. Monitored natural attenuation may not be a suitable sole remedy in some fractured rock settings where risk to human or ecological health is severe.

The occurrence of natural attenuation processes within fractured rock systems, and their demonstration at a particular site, is not necessarily sufficient for the use of monitored natural attenuation as a sole remedy (e.g., EPA, 1999a). Reaction processes (i.e., biodegradation and chemical) that reduce contaminant within a system and that are fundamental to achieving a steady-state plume configuration are potentially less common and likely less strong in the majority of fractured rock settings as compared to porous media settings (Chapelle et al., 2012; Lima et al., 2012). Organic carbon, nutrients, and electron donors are often found in insufficient quantities in fractured rock environments to promote and promulgate biodegradation, which results in less dominant reaction processes and longer and continually expanding plumes. The reduced magnitude of these processes in fractured rock is often masked by matrix diffusion over relatively long timescales (as compared to porous media) imparting an artificially increased importance to reactive processes.

A high level of confidence in monitoring is required to effectively use MNA. Principles of MNA require a suitable, representative, conservative, and fit-for-purpose monitoring regime (Wiedermeier et al., 1998). The level of uncertainty in monitoring programs at many fractured rock sites may exceed that allowable for MNA to be a protective approach, and the capacity to reduce that uncertainty at some sites may be beyond that of the responsible party. It is entirely conceivable that a line of evidence within the MNA framework may show that the plume has dissipated or depleted, where in fact the monitoring infrastructure is not properly positioned within the flow paths and is no longer monitoring the plume. The level of characterization required for acceptable uncertainty in the delineation of the plume in fractured rock systems is likely orders of magnitude above that required in most unconsolidated systems, and application of MNA will likely require demonstration of adequate characterization of fate and transport processes and pathways (e.g., EPA, 2001). Furthermore, natural attenuation may not reduce contaminants to levels sufficiently or quickly enough to avoid unacceptable risk to health, as demonstrated at the Cayuga Springs Superfund site (EPA, 2012).

Concept 8. Current metrics are not well suited for remedial objectives at fractured rock sites.

NRC (2013) proposed that better metrics for demonstrating remediation progress are needed that are based on quantifiable, transparent metrics of remedial performance and human health reduction rather than regulatory milestones. Contaminants typically cannot be removed from an interconnected fracture system to meet MCL cleanup goals unless some percentage of the contaminant mass in the matrix is removed. The contaminant flux through back diffusion needs to be reduced to a point where concentration-based standards in groundwater sampled from monitoring wells (which is, in most cases, derived primarily from fracture flow) are met. A more suitable metric for remedial performance may be the reduction in concentration of contaminants within the matrix itself, coupled with demonstration of contaminant source removal (e.g., DNAPL) in the fractures. Such a matrix cleanup goal requires consideration of a coupled system of transmissive fractures and low permeability matrix (Rodriguez and Kueper, 2013).

Given the heterogeneous nature of transmissivity and the often complex flow patterns in fractured rock, the use of point measurements of concentration in monitoring wells as performance objectives is potentially flawed. As described in Chapter 5, monitoring wells in fractured rock may intersect a single, highly contaminated flowing fracture, and the measured concentration may not represent the average system concentration. The value, therefore, is not suitable for comparison to a risk-based criterion that assumes an exposure pathway (be it through dermal contact, ingestion, or as a source for vapors) comprised of only that water. Measurement of contaminant flux across a control plane is potentially a more applicable metric for determining the risk to human and environmental health in fractured rock environments.

Concept 9. At some point, active remediation is no longer appropriate, and long-term management should be formally implemented.

NRC (2013) has suggested that the transition to long-term management should be recognized formally as the point when further active remediation results in little or no decrease in contaminant concentration and when the unit cost of remediation increases much faster than reductions in contaminant concentrations. Under this approach, the transition point to long-term management (i.e., monitored natural attenuation) is likely to occur far earlier in fractured rock than in unconsolidated systems. However, both the suitability and the acceptability of this occurrence are less clear in fractured rock settings. For example, based on the predictions of Parker et al. (2010), any active remediation of the initial source mass (note that the initial source contaminant is representative of the classic unconsolidated source and does not consider back diffusion from the matrix porosity within the "plume" as a source) could be reasonably argued to have effectively little change on contaminant concentrations and thus not be required. Such an approach is not suitable as a default position for remediation at fractured rock sites, but it may be suitable as a considered position at some sites.

THE PATH FORWARD

The differential between the state of practice and state of the art in fractured rock investigation and remediation is greater than in unconsolidated porous media. The reasons for this are numerous and include less experience among practitioners, the greater difficulty of investigation in fractured rock, a much shorter history of detailed scientific understanding, and a residual impression among most stakeholders that the RI/FS process between fractured rock and unconsolidated porous media is similar, and so there is a high level of transferability of approaches and outcomes between the two. Reducing this differential in remediation will require new and increased effort in several areas:

1. Recognition and understanding of the differences between fractured rock and unconsolidated porous media.
2. Collaborative efforts between researchers, practitioners, and regulators to develop meaningful and obtainable outcome frameworks for remediation.

3. Scientifically rigorous and peer-documented demonstrations of various remediation technologies in different geologic settings.
4. Increased strategic characterization prior to remediation design utilizing appropriate techniques.
5. Development of monitoring strategies based on valid conceptual models, targeting appropriate locations and times, and specific parameters whose changes are related to remediation performance success or failure.

Improving practice will require aligning the goals and resources of researchers, funding agencies, regulators, site owners, and practitioners. Reduced research funding for fractured rock remediation studies leaves few options for implementing and managing a program or process to increase knowledge and improve practice. Without a governing body to focus and direct efforts on this process, and without inclusion of the oil and gas industry in the process, the knowledge gaps will continue to be a significant impediment to progress on effective management of the nation's contamination issues in fractured rock.

7

Decision Making

As discussed in previous chapters, the characterization, modeling, monitoring, and remediation of the fractured rock environment is challenging because of (1) the complexities of the mechanical, chemical, and biological environments of fractured rock, and (2) the spatial and temporal variability of the processes that affect fluid flow, contaminant migration, and contaminant storage and degradation. The quantity and quality of information available for fractured rock sites varies, making the development of a flexible but robust engineering decision framework a challenge.

A decision-making framework for engineering or operating engineered facilities in fractured rock sites needs to be adaptable to a wide range of conditions and the most critical engineering questions. Historically, engineering goals such as remediation have been viewed as endpoints or terminal processes in linear activities. However, earlier chapters in this report demonstrate that engineering goals and solutions can be refined and made more effective if informed regularly by new characterization, modeling, and monitoring data. This implies that a shift from a linear to a cyclic decision-making framework will result in more effective and efficient engineering.

The importance of understanding uncertainties in characterization, modeling, monitoring, and remediation of fractured rock is stressed throughout this report. Much decision making related to engineering a site—whether for future waste disposal, remediation of existing contamination, or performance monitoring of any manner of infrastructure—is based on some form site conceptual model and needs to include consideration of uncertainty. Indeed, the decision-making framework should be designed to consider and, if possible, reduce uncertainties to attain effective and efficient engineering solutions. A classification structure for uncertainties is summarized in Box 7.1.

The consistent and systematic consideration of uncertainties is critical in fractured rock engineering. Decision making under uncertainty has a long history in business management (Pratt et al., 1965; Raiffa, 1968; Staël von Holstein, 1974). The underlying procedure is schematically captured in Figure 7.1. Each phase of the cycle outlined in this figure is directly mappable to the engineering solutions described earlier in this report. By the scheme depicted in Figure 7.1, information collection occurs early and throughout the process (i.e., characterization), but more information may be gathered as warranted (i.e., more characterization and monitoring). Deterministic and probabilistic modeling can be applied to both hydrostructural (conceptual) model development and numerical modeling (simulation). The process of decision making has only been discussed implicitly and without explanation earlier in this report. This chapter first describes ideas and concepts for decision making in the related fields of rock mechanics and geotechnical engineering. It is then shown how these concepts could be applied to the characterization, modeling, monitoring, and remediation of fractured rock. The concept of updating is also discussed. Updating describes the various processes that result in change, such as the acquisition of additional information or the application of remediation strategies.

DECISION MAKING AND UPDATING IN THE CONTEXT OF THIS REPORT

Models depict outcomes such as the prediction that a particular concentration of a contaminant reaches some level in the rock mass. These outcomes are rarely deterministically expressed in fractured rock settings. Outcomes from deterministic models can be compared to regulatory limits, model outcomes given different input parameters, or outcomes of another model. In almost all cases, however, a probabilistic model can provide valuable insights regarding the relationship between uncertainties and various model predictions. Simple or complex decisions can then be made, for example, on the choice of a particular model, the need for more model input parameters, or the choice of specific remediation strategies. Making decisions based on model outcome underlies even the most complex process.

BOX 7.1 Classification of Uncertainties

There is a need in geotechnical engineering to address and reduce risks associated with difficult-to-observe subsurface conditions. Uncertainties in geotechnical engineering can be classified into five categories (Einstein and Baecher, 1982):

1. Innate spatial and temporal variability
2. Errors induced by testing and estimation of engineering properties
3. Model uncertainty
4. Load uncertainties
5. Omissions

Uncertainty may reflect randomness (i.e., aleatory uncertainties) or lack of knowledge (i.e., epistemic uncertainties) (Christian et al., 1994). Causes of uncertainty may be the individual and combined effects of the five categories above or other factors. Further discussion and specific examples can be found in Lacasse et al. (2012).

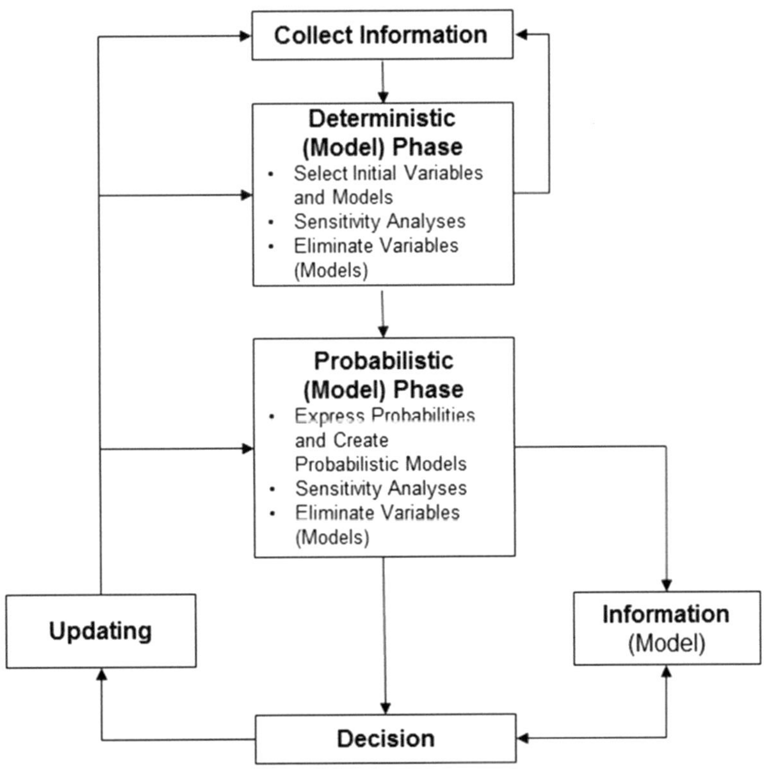

FIGURE 7.1 A decision analysis cycle for geotechnical engineering based on a business modeling approach. SOURCES: Modified from Einstein, 2002, after Staël von Holstein, 1974.

Updating

Action logically follows a decision. In the context of decision making under uncertainty, this action is called *updating* (see Figure 7.1). Formal procedures, mostly based on Bayes' updating (Ang and Tang, 1975), exist and have been widely applied (e.g., Einstein, 1988). Collecting additional information changes (updates) model input parameters and may result in different decisions being made. Choosing a different model or modifying a model based on new information are both forms of updating the models. These may result in changed model output and altered decisions. Because remediation activities change environmental conditions, they are also a form of updating. Information collection itself is also subject to uncertainties and so decisions are made regarding the need to collect information. Literature on how to apply the information model, and the concepts and application of value of information (specifically, the value of sample information and the value of prefect information) in the context of geotechnical and geoenvironmental engineering exists (see, e.g., Einstein et al., 1978; Cardiff et al., 2010; Neuman et al., 2012; Sousa et al., 2014).

The information collection and the models that are part of the decision cycle have been extensively covered earlier in this report. The application of the observational method to the decision process is discussed below.

OBSERVATIONAL METHOD IN GEOTECHNICAL ENGINEERING

The observational method in geotechnical engineering is a process that involves investigation, characterization, design, construction, and management based on the most probable conditions expected at a site rather than the worst-case scenario. This section summarizes the observational method in geotechnical engineering for those unfamiliar with it. Because it is recognized that much is learned about a geotechnical project during the project cycle, pre-determined modifications to design, construction, or operation can be made at any time based on review of observations. The observational method is intended to minimize project costs—without sacrificing project safety—by allowing for flexibility in decision making based on updated information.

The term *observational method* appears to have been proposed by Terzaghi (1961), but he refers to a very early application at the beginning of the 20th century. Philosophically, one can relate the observational approach to what was proposed by the geologist T. C. Chamberlin (1897), namely, the multiple working hypotheses. Chamberlin dealt with the uncertainty in geologic interpretations, similar to what the geotechnical observational method attempts to do. This method, which follows the principle shown in Figure 7.2, is well established, although not always used. Figures 7.1 and 7.2 have strong similarities, in particular the feedback cycle in Figure 7.2, which corresponds to the updating in Figure 7.1.

The essence of Figure 7.2 is that a number of design options are developed, each of which includes an associated performance prediction. This predicted performance is then compared to the observed performance, and a decision is made to either maintain the current design or use another design (i.e., update the decision). All of this reflects the fact that geologic and geotechnical conditions are uncertain and that there is a need to react flexibly to new information. This decision-making method is used in other applications, such as the recommended approach for design of carbon dioxide injection wells for the purpose of sequestering carbon. In such applications, injector wells are placed, observed, and operated prior to collection of performance data (Cameron and Durlofsky, 2014). Model predictions are then updated, and well settings may be altered in response to performance data collected. An iterative approach that includes quantifying uncertainties is also suggested for the decision methodology for both mined repositories (Meacham, 2011) and deep borehole disposal of nuclear waste (Arnold et al., 2012).

Expanding beyond Terzaghi's description of the method, Peck (1969) proposed a detailed step-wise approach (see Box 7.2). One consequence of the observational method, however, is that it can be difficult to bid and award a contract in a fixed cost (time) framework. This is because the approach recognizes the appropriateness of different designs and outcomes depending on observed site conditions. Possible ways to formulate contracts for an adaptable approach have been discussed (e.g., Stasiewicz, 1981; Patel et al., 2007), but the fixed cost (time) environment traditional in public works contracts makes the observational approach difficult to apply. The observational method is often not employed fully or is ignored because of fixed regulatory requirements. An application of the observational method in fractured rock that could inform infrastructure management and site remediation approaches is in the context of the new Austrian tunneling method (NATM). Examples of the application of the NATM are Tentschert et al. (2003) and Golser and Schubert (2003). Other recent examples of application of the observational method in the fractured rock environment include Bäckblom and Öhberg (2002), Hernqvist et al. (2012), and Olofsson et al. (2014). The adaptability and flexibility of the observational approach is not only well suited for geotechnical engineering but is consistent with approaches to fractured rock site remediation and management as discussed in Chapter 6, and for geoenvironmental problems more generally.

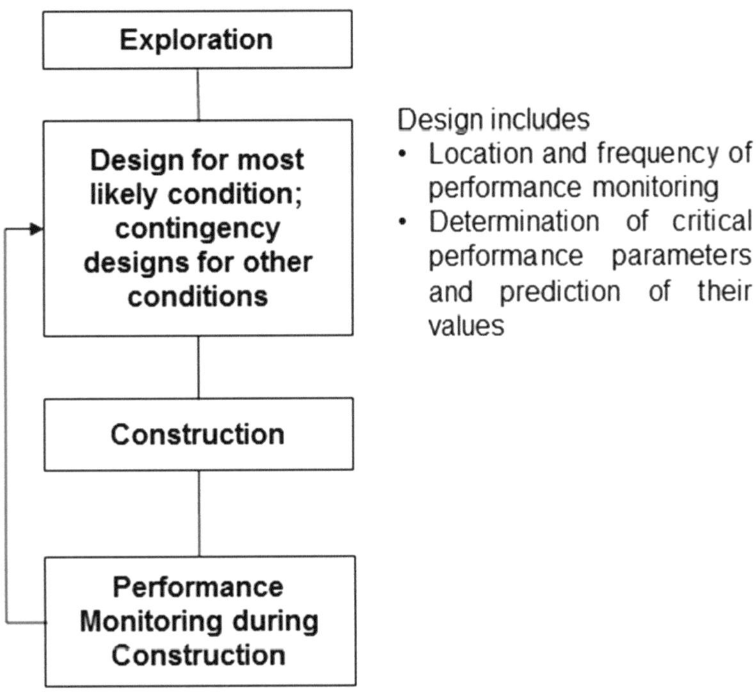

FIGURE 7.2 Schematic representation of the observational method. Knowledge of design parameters is updated through monitoring and feedback into design. SOURCE: Modified from Einstein, 1995.

BOX 7.2 Stepwise Approach to the Observational Method

Peck (1969, p. 173) described the steps necessary for a complete application of the observational method during construction in soils. Many of those steps, paraphrased below, are also directly applicable to engineering in the fractured rock environment.

1. Explore sufficiently to establish at least the general—but not necessarily detailed—nature, pattern, and properties of the soil deposits.
2. Assess the most probable conditions, and the most unfavorable conceivable deviations from those conditions, based largely on geology.
3. Develop designs based on a working hypothesis of behaviors anticipated under the most probable conditions.
4. Select quantities to be observed as construction proceeds, and calculate their anticipated values on the basis of the working hypothesis.
5. Calculate values of the same quantities under the most unfavorable conditions compatible with the available data concerning the subsurface conditions.
6. Select for every foreseeable significant deviation of the observational findings from those predicted on the basis of the working hypothesis in advance of a course of action or modification of design.
7. Measure quantities to be observed and evaluate actual conditions.
8. Modify design to suit actual conditions.

EVOLVING THE OBSERVATIONAL METHOD TO GEOENVIRONMENTAL ENGINEERING

Geoenvironmental engineering decisions are often driven by compliance-based requirements. For example, decision processes employed routinely by many engineers and scientists for cleanup of contaminated sites in the United States are heavily influenced by requirements of the U.S. Environmental Protection Agency (EPA) under the Comprehensive Environmental Response, Compensation, and Liability Act, commonly referred to as Superfund (EPA, 1999b). The highly linear decision process is described in EPA's guidelines and involves

1. Remedial investigations and feasibility studies (site characterization, treatability studies, evaluation of alternatives),
2. A remedy selection process (identification of preferred remediation scheme),
3. Development of a proposed plan,
4. Opportunity for public comment,
5. Remedy selection,
6. Announcement of a record of decision,
7. Remedy implementation, and
8. Long-term maintenance.

This linear decision-making process provides little opportunity to learn from new information (i.e., update through feedback loops) and iteration of any step once a decision is made and recorded as a public document (typically by a judge at the end of a lengthy, sometimes contentious, and expensive adjudication process). This process can be effective for relatively simple sites that can be well characterized with acceptable levels of uncertainty (e.g., through pump tests), and when conceptual site models can be analyzed readily with available numerical techniques to predict the fate and transport of contaminants. A linear process facilitates quick deployment of remediation designs, but new information seldom results in radical modification of conceptual and numerical models or remediation plans.

Most fractured rock sites, however, require greater levels of characterization and are burdened with higher levels of uncertainty. Under these circumstances, engineering solutions rarely are arrived at using the linear processes favored under the current regulatory environment. Conceptual and numerical models for fractured rock sites require more variables to be quantified, more assumptions to be made, and contain greater uncertainty in both quantification and qualification. The best models tend to be those that are modified as new information is gathered during characterization, monitoring, and remediation—methods far more consistent with the observational approach. This approach to model development can lead to better understanding of a site and the most effective site management approaches. However, the current regulatory environment does not accommodate these approaches.

Application of the observational method is relevant particularly when performance depends on complex coupled processes at a variety of temporal and spatial scales, for example, the isolation of nuclear waste in fractured rock settings. As noted by Yow and Hunt (2002, pp. 145–146):

> we measure most rock properties and processes in experiments and tests that are orders of magnitude smaller and of much shorter duration than engineering projects. Together with the complications of site heterogeneity and parameter variability, this makes it difficult to predict and extrapolate the effects of coupled processes to larger scales and over time.

The observational method is embedded in adaptive site management (NRC, 2003) and in long-term management when traditional site remediation technologies prove ineffective (NRC, 2013b). In all cases, the decision-making process needs to be applied throughout the lifecycle of the engineered facility, which may extend over millennia, and be used to continuously reassess engineering goals, available and needed information and associated uncertainties, and appropriateness of engineering decisions. If new information cannot reduce uncertainties to acceptable levels, or if monitoring indicates that implemented decisions are not producing predicted results, then it is necessary to question the appropriateness of the original engineering goals.

The decision process outlined in Figures 7.1 and 7.2 is reframed in Figure 7.3 to reflect the vocabulary and processes central to this report, and Box 7.3 provides a theoretical application of the observational method. It is intended to guide rather than prescribe decision making under almost any geoenvironmental engineering circumstance and for most temporal and spatial scales.

Effective geoenvironmental engineering in fractured rocks requires iterative and adaptive processes to minimize physical risks to health and environment. The observational method works around the inescapable fact that needed information about a fractured rock site generally is gained only once the project advances. A proper decision-making approach includes multiple decision points at which available information is assessed, decisions can be improved based on an increasingly accurate site model and more reliable information, and project goals are met, sustained, and re-evaluated when necessary. Decision making within an observational approach is uniquely adapted to address these needs.

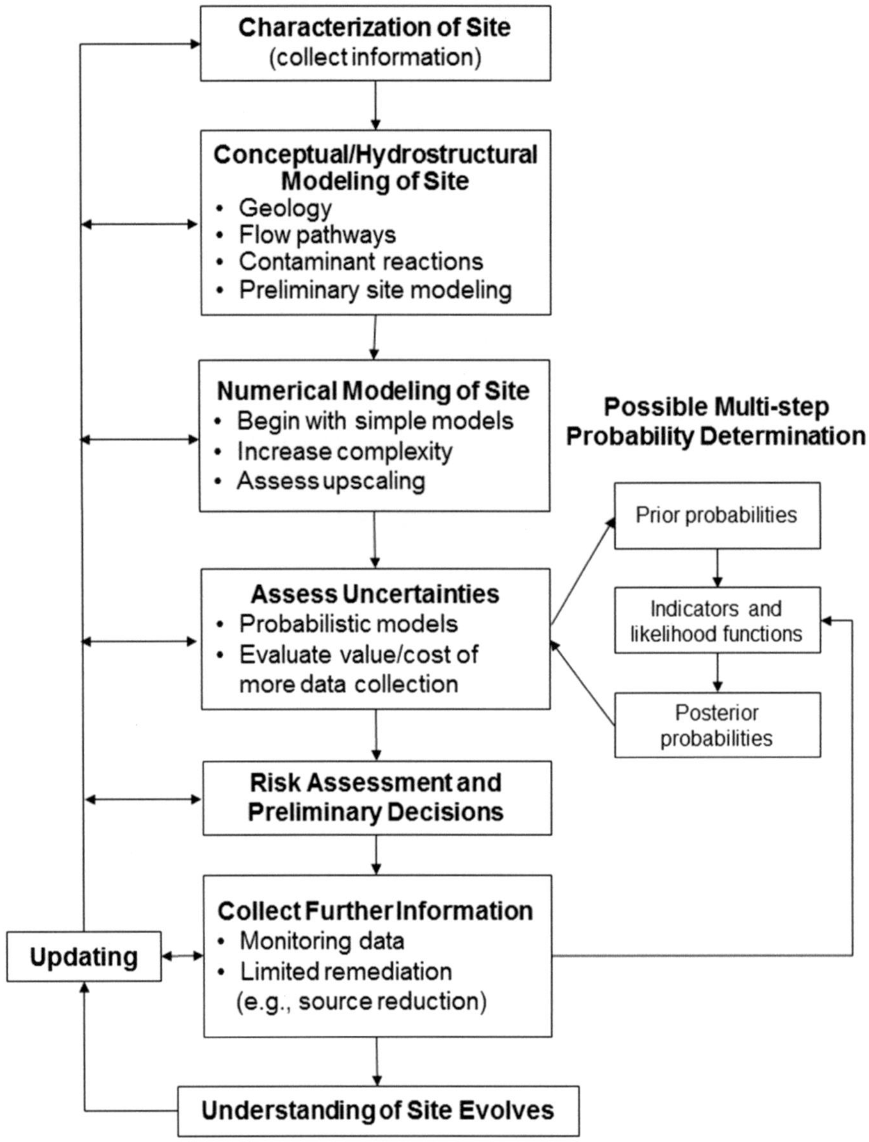

FIGURE 7.3 Adaptation of an observational approach to engineering at a fractured rock site.

BOX 7.3 Theoretical Application of the Observational Method in Geoenvironmental Performance Monitoring

Geoenvironmental performance monitoring of a storage facility will be based on some sort of engineering goal, for example, to detect leaks. The observational method requires comparison of actual performance observations with expected design performance, deviation from which triggers re-evaluation of available information to assess causes and impacts. Additional information needs will be determined and prioritized, and data will be collected based on available resources. The compiled data then inform predictive modeling conducted to identify the cause of the deviation. Modeling results may inform updates to the site conceptual model. Additional characterization may be warranted until concept and numerical models have acceptable levels of uncertainty and reliability. If model results continue to represent unacceptable levels of uncertainty and reliability, then it may be decided that the original engineering goals are unrealistic and in need of modification. Engineering decisions can then be made to address the cause of the deviation (i.e., to make repairs) and to address any impacts (i.e., remediate contaminated groundwater). These engineering decisions, in turn, will be evaluated for acceptable levels of uncertainty or reliability before implementation and further monitoring.

8
Synthesis of Recommendations

Accounting for important fractured rock features and processes is necessary and possible for characterization, modeling, monitoring, and remediation. Tools to analyze and characterize fractured rock have developed rapidly in the past 20 years and can provide a basis to solve or avoid problems related to contaminants in fractured rock near surface and at depth. Discrete groundwater flow pathways, multiple porosities, and the hydraulic and biogeochemical communication between different characteristic porosities can often be accounted for at reasonable levels of uncertainty and reliability if the right questions about them are asked. Ineffective engineering approaches may then be avoided, and realistic solutions can be designed within technical, logistical, and economic constraints.

This chapter synthesizes the recommendations found throughout this report and identifies particularly promising areas for research and development related to fractured rock systems. These recommendations are all applicable in some degree to practicing engineers, regulators, infrastructure owner and operators, and researchers. They are intended to improve engineering practice given today's tools and knowledge, guide a framework for research that can inform future practice, and inform the decision maker who will benefit from a more realistic understanding of current capacities and limitations.

AN INTEGRATED APPROACH

An integrated approach to engineering in the fractured rock environment is essential, but all too often engineering decisions are made by engineers and infrastructure managers without an integrated perspective. Integrating geologic, geophysical, hydrodynamic, and geomechanical information is essential to understanding fractured rock sites. Geologic features (e.g., fractures, discrete beds, sealing faults, rock matrix) that dominate fluid flow, transport pathways, and storage need to be understood to characterize fractured rock sites correctly and predict fluid and contaminant fate and transport.

Flow and transport are influenced by issues of scale for both fractures (μm to km) and rock matrix porosities (μm to cm) that, in turn, affect advective, diffusive, chemical, biological, capillary, and filtering processes over different timescales. Changes in flow and transport behavior can result from the combined effects of these processes (e.g., dissolution and precipitation can change preferential flow pathways) and therefore need to be considered from the earliest stages of analysis. Because a few fractures are likely to control the majority of groundwater flow, minor perturbations in fracture surfaces and connectivity resulting from geologic and biogeochemical processes are important in contaminant transport and distribution. This does not imply that large and complex geologic and biogeochemical analyses are always necessary—but does require that such issues be recognized and evaluated at appropriate qualitative or quantitative levels.

Recommendation 1. Take an interdisciplinary approach to engineering in fractured rock and use site geologic, geophysical, geomechanical, hydrologic, and biogeochemical information to conceptualize transport pathways, storage porosities, fate-and-transport mechanisms, and the coupled processes that control rock fracture-matrix interactions.

To facilitate this recommendation, engineers who are characterizing, modeling, monitoring, or managing engineering project sites need to

A. Delineate at the onset the processes that can affect contaminant transport, storage, and transfer between the rock matrix and fractures, particularly over the life of a remediation project and follow-on monitoring. Determine the importance and potential for hydrologic, thermal, biological, chemical, and mechanical processes—and the coupling between those processes and their resulting feedback effects—in both fracture and matrix porosities.
B. Use available geologic, hydrogeologic, and geophysical information to conceptualize flow pathways, fluid and contaminant storage, alteration or attenuation through geochemical reactions, and degradation via biological processes.
C. Conceptualize fractured rock hydrogeology by first assuming dual porosity/single permeability discrete fracture network (DFN) (i.e., primary storage of contaminant in the rock matrix and primary flow of fluid in the fracture network), and then modify to simpler (single effective porosity and permeability equivalent porous media) or more complex (dual porosity/dual permeability hybrid DFN) assumptions when justified by site-specific evidence.
D. Recognize the limitations of advection-dispersion-diffusion approaches to solute transport solutions, and consider which approach is most appropriate. An advection-dispersion model may be good at fitting transport measurements at a specific scale, but the anticipated groundwater velocity variability in fractured rock aquifers is inconsistent with the underpinnings of the advection-dispersion model. Therefore, the model parameters obtained in that fitting process often are not appropriate for use at other scales. Consider, for example, the potential for gravity-driven flow, heterogeneous fingering, and rock matrix imbibition processes for multiphase flow.
E. Recognize the frequently dominant role of discrete fractures in flow and transport. Distinguish hydraulically active fractures from those that provide only fluid storage through hydraulic testing, passive monitoring of pressures and chemistry, and geophysical and geomechanical measurements. Design monitoring systems to refine site conceptual models, account for discrete flow and transport, and aid engineering decision making. Do not rely solely on integrated measures across large sections such as long, open boreholes. Recognize how fractures and rock matrix properties and processes change with depth, stress, and geochemical and geobiological conditions.

INTERACTIONS BETWEEN ROCK MATRIX AND FRACTURES

Understanding the interactions between rock matrix and rock fractures is critical to understanding contaminant fate and transport. The potential for storage of contaminants in the rock matrix, and the slow diffusive transfer from the matrix to flowing fluid in adjacent fractures, can result in remediation time frames of decades or centuries. It is important to understand such subsurface interactions at the onset of infrastructure development to inform appropriate mitigation measures. Engineers responsible for site selection, characterization, project permitting and licensing, project construction, and infrastructure management need to be aware of the potential of contaminant release and post-release behavior—not only during the early stages of project development, but also for the lifecycle of the infrastructure. Failure to understand storage and release mechanisms can lead to gross miscalculation of subsurface contaminant distribution and effectiveness of remedial measures (e.g., pump and treat).

Recommendation 2. Estimate the potential for contaminant to be transported into, stored in, and transported back out of rock matrix over time.

A. Recognize that fractures and fracture networks can be discrete flow pathways or flow barriers or provide only storage porosity. Develop and implement strategies to characterize mobile porosity geometries (e.g., structurally controlled flow pathways such as those defined by faults and karst features) and immobile porosities (e.g., low-permeability fractures, fractures not connected to other fractures, and rock matrix). Recognize uncertainties in geometry of heterogeneities, pathways, and barriers. Conceptualize boundary conditions, sources, sinks, and disturbed rock zones.
B. Determine how to quantify and differentiate contaminant in mobile porosities from that in immobile porosities (generally rock matrix porosities) at various locations (e.g., source zones, impact zones, upgradient/downgradient zones).
C. Evaluate the contaminant stored in and transferred between rock fractures and matrices, recognizing the various processes that might occur (e.g., advection, diffusion, sorption, biodegradation, filtration, and capillary processes). Consider the effect of fracture and rock matrix void sizes on capillary and wetting processes, bioactivity, and multiphase flow.
D. Develop methods to better delineate the distribution of non-aqueous phase liquids within mobile porosity. Apply advances of the past 20 years (e.g., fracture image logging, borehole flow logging, partitioning tracers, and wireline geochemical monitoring) to characterize conductive discrete features. As appropriate, develop and confirm hypotheses concerning transport pathways and geometries using geologic, geomechanical, and geophysical techniques.

PROCESSES AND COUPLED PROCESSES

Chemical, biological, thermal, mechanical, and hydraulic processes can individually and in combination change how fluids containing contaminants, migrating fines, and colloids interact with the rock matrix and fracture infillings and coatings over time. The importance of coupled processes in fate and transport is not always recognized or addressed by engineers, infrastructure managers, and regulators. Furthermore, the timescales of changes associated with some processes may be longer or shorter than the life of the engineering project.

Recommendation 3. Characterize chemical, biological, thermal, mechanical, and hydraulic processes, their interactions, and the conditions that can lead to their coupling to better understand transport through discrete fractures and contaminant transfer between fractures and rock matrix.

A. Consider the possibility of flow localization within fractures, channels within fracture planes and intersections, the role of hydro-mechanical coupling on changes in fracture and channel geometry, and the effect of these changes on advection, dispersion, and diffusion in the interpretation and prediction of transport rates.

B. Evaluate the potential role of mineral fines to promote contaminant transport or clogging.
C. Evaluate fluid properties (e.g., miscibility and mutual solubility, density, viscosity, and acidity) and reactivity with rock or infilling material. Assess timescales of fluid-mineral reactions to determine their impact in terms of the life of the engineered project.
D. Recognize that immiscible fluids behave differently from water and understand the pore size-dependent capillarity reductions in permeability that result from phase interference.
E. Consider the role of bioactivity—and limiting environmental factors (e.g., pore size and availability of nutrients)—on dissolution, precipitation, biofilm formation, and contaminant transformation.
F. If thermal changes are anticipated, then assess subsurface heat conduction and temperature changes under static- and advective-flow conditions in consideration of flow localization.
G. Consider differences in fracture and rock matrix properties, and mechanical, chemical, and biological conditions with depth, particularly with changes in stress, temperature, and groundwater chemistry.

Biological Processes

Current remediation technologies are limited in their ability to reduce contaminant concentrations in fractured rock to desired levels within short time frames. Because of the potential longevity of contaminants retained within the rock matrix, time frames associated with reaching water quality standards may range from decades to millennia. Biological processes are known to be important in certain situations, but the magnitude of that importance in contaminant fate, transport, and remediation is not quantifiable in a general way. Their importance can be quantified for a specific situation given enough information. Characterization of biological communities and activities in fractured rock will allow for better characterization and predication of fluid and contaminant fate and transport. It is likely that biological processes will be relied on to reduce and control the further spreading of in situ contamination at most contaminated rock sites because no other alternative is likely to prove practical in many situations. The single biggest research need is, therefore, to understand better biological processes at fractured rock sites, and especially rock at great depth where there is a dearth of information about biological processes.

Improvements of molecular tools in the past two decades have advanced knowledge of biological processes. Genome sequencing at the individual cell level, for example, is now possible. However, subsurface microbial characterization is still hampered by difficulties studying in situ biological systems that allow the processes and rates of degradation to be understood. Reliable, reproducible, quantitative, and statistically valid experimental information on the spatial and temporal dynamics of biological communities is needed so that valid, quantitative assessments can be made of the biodegradation of contaminants in fractured rock. Application of advanced molecular tools (e.g., advanced sampling techniques and high-throughput metagenomics/metatranscriptomics technologies) coupled with computational modeling should make it possible to systematically address fundamental microbial questions not possible previously because of difficulties sampling and cultivating microorganisms from fractured rock environments.

Recommendation 4. Expand research to define and quantify microbial influences on fluid and contaminant fate and transport in fractured rock over timescales relevant to contaminant remediation processes.

Specific research topics could include

A. The extent of phylogenetic, genetic, and functional diversity of microbial communities in fractured rock environments;
B. The influence of microbial populations on hydraulic and hydrogeochemical characteristics of fractured rock environments;
C. Changes of phylogenetic and functional structures in microbial communities, and associated relationships with hydrogeochemistry and the anoxic, high-salinity, high-temperature, and high-stress conditions of the deep environment across various spatial and temporal scales; and
D. Microbial community response to environmental perturbations within fractured rock environments.

CHARACTERIZATION TECHNIQUES AND TOOLS

Site characterization methods need improvement. At present, site characterization includes application of tools and techniques for geologic, geophysical, geochemical, hydraulic, biological, and geomechanical observation, testing, and sampling. However, most of these methods are limited in terms of their ability to map fracture characteristics at appropriate scales. Borehole sampling and testing tend to be expensive and provide limited information on a relatively small scale, but surface-based methods such as seismic may not have appropriate resolution to map fractures directly. Cross-hole tracer testing is frequently needed to understand large-scale issues such as fracture connectivity and flow dimension, but it does require boreholes—often a limiting factor. Engineers need to better understand the benefits and limitations of characterization data from various sources, and project managers and regulators need to understand what constitutes an honest appraisal of the data. All need to remain aware of innovations in site characterization technologies and be willing to incorporate those developments into practice, particularly borehole- and surface-based technologies from the petroleum and mining industries for characterizing rock at depth.

Supporting text for the recommendation below includes contributions that the research community could make that could be supported by the industry and regulatory communities.

Recommendation 5. Improve characterization and monitoring through new and expanded research in surface- and borehole-based geologic, geophysical, geochemical, hydraulic, biologic, and geomechanical technologies.

Improve characterization and monitoring by advancing, for example

A. Research on technologies such as geophysical tools, innovative tracer tests, and flexible liners capable of measuring contaminants in situ to identify, characterize, and monitor the volumes, spatial distribution, and transfer of contaminants between rock matrix and fractures;
B. Joint inversion methodologies for fracture and process parameterization;
C. Seismic, microseismic, and hydraulic tomography, hydraulic interference, and tracer testing techniques that allow for characterizing flow paths at a range of scales, depths, and in situ conditions;

D. Fracture mapping techniques (e.g., microseismic, thermal, electrical resistivity, and geochemical) to identify conductivity and flow in fractures, calculate in situ gradients, and determine flow directions at a range of scales, depths, and in situ conditions;
E. Techniques for tracking tracers or certain contaminants, including, for example, microseismic, electrical, nuclear magnetic resonance, and radar geophysical techniques; and
F. Development of high-throughput molecular techniques for understanding fractured rock biology and bioremediation possibilities.

MODELING

Current numerical model applications used by practicing engineers and researchers often do not account for discrete fracture flow pathways and fracture-matrix interactions. Budget, data, personnel, and time frequently constrain the choice of modeling methods, and practice tends to focus on building and calibrating a single porosity, equivalent continuum numerical model of flow and transport. Although it is frequently possible to calibrate such models to measurements, their predictive powers can be limited because they do not incorporate discrete features and fracture-matrix interactions. They also inform in a limited way the conceptual and parametric sensitivities and uncertainties associated with model assumptions. A variety of powerful preliminary scoping calculation methods can be used by practicing engineers and researchers to quantify the importance of specific processes and assumptions. Regulators need to be aware of the appropriate use of conceptual and numerical modeling methods to inform their decisions about the suitability of site engineering design and operations.

Recommendation 6. Develop appropriate hydrostructural conceptual models for fracture and rock matrix geometries and properties, and perform preliminary calculations (e.g., analytic or simple numerical) to better inform and allocate resources for site characterization, modeling, and remediation.

A. Follow a systematic and well-documented hydrostructural conceptual model development approach so that underlying assumptions and simplifications are well understood, especially in the absence of supporting data. An appropriate hydrostructural model includes definitive fracture and matrix porosities, spatial and temporal uncertainties, and variabilities for a site.
B. Begin with generic geologic and geomechanical conceptual models and parameterizations informed but not limited by site-specific direct measurements.
C. Develop a broad, semi-quantitative understanding of important processes and parameters, refined with simplified scoping calculations that define and assess
   - alternative multi-porosity fracture flow and transport conceptual models;
   - model uncertainties;
   - the appropriateness of simplified versus more complex modeling approaches; and
   - how proposed site characterization, analysis, and modeling activities may reduce uncertainties and improve engineering decisions.
D. Quantitatively integrate field data into models through model analysis tools such as parameter estimation, sensitivity analysis, and uncertainty quantification.

If numerical models do not represent adequately the hydrostructural conceptual models developed for a site, as is common in current practice, then they will inadequately predict site response to in situ conditions and remedial actions, including contaminant transport and retention. For near-surface applications, and those at depths, it is essential to incorporate appropriate hydrostructural features (e.g., discrete fracture pathways and matrix storage) into numerical models—even in a simplified manner.

As described earlier, single porosity continuum-based approaches inadequately represent the effects of both fractures and rock matrix porosities on contaminant pathways and storage. Discrete fracture models that properly simulate flow, transport, and coupled processes can be difficult and complex to develop, but they may be required for adequate projections of contaminant fate and transport.

Recommendation 7. Base numerical models on an appropriate hydrostructural model, ensuring that simplification and upscaling of the hydrostructural model maintain those features, properties, and processes that dominate contaminant transport, fate, and storage. Evaluate impacts of uncertainties introduced by simplification and upscaling.

During analysis and modeling

A. Understand the limitations and advantages of numerical models and analysis tools to represent the hydrostructural model and fractured rock environment. Choose tools that synthesize data, confirm or refute the validity of the conceptual model, and quantify the range and uncertainties of expected system behaviors.
B. Consider the scale of the problem of interest relative to the scales of the rock fractures and matrices. Apply equivalent continuum models only when the problem scale is much greater than the scale of the fractured rock mass. Equivalent continuum and fracture network analysis upscaling techniques can produce anisotropic and heterogeneous continua, but not necessarily correct discrete pathways.
C. Consider
    - The implications of the use of simplified and effective medium modeling approaches;
    - The need for risk- and uncertainty-based modeling approaches, even where calibrated and conditioned models match available measurements;
    - Approaches capable of addressing multiple porosities, discrete pathways, spatial heterogeneity, and decision making under uncertainty;
    - Non-deterministic modeling techniques applicable when there are no current means to identify groundwater paths and response functions; and
    - Modeling at multiple scales, for example near well and over the full pathway from depth to surface of boreholes.

REMEDIATION AND MONITORING

Expectations related to site remediation often are unrealistic. The long-term goal of remediation at most hazardous waste sites in the United States is to meet drinking water requirements associated with Maximum Contaminant Levels (MCLs) at monitoring points or at actual or potential receptors. However, meeting MCLs at fractured rock sites can take decades or centuries. Because long time frames for remediation are becoming more acceptable to U.S. regulatory bodies, alternate approaches to remediation, particularly at fractured rock sites, are becoming more accepted.

Much information regarding innovative contaminated site remediation is available through, for example, an Environmental Protection Agency–sponsored website.[1] Although information about fractured rock site remediation has been published, it is less widely available. Lessons learned from remediation and monitoring activities in the fractured rock setting are not well documented. As a result, there is risk of duplicated effort and wasted resources in practice. A significant impediment to the remediation of fractured rock settings is the lack of common framework, understanding, or expectations regarding objectives, assessment, and realistic remediation endpoints.

Recommendation 8. Develop and communicate realistic expectations related to remediation effectiveness through realistic goal setting and through explicit consideration of uncertainties in design, realistic use of natural attenuation, comprehensive monitoring programs, and dissemination of performance data to the technical community.

Practitioners and regulators should

A. Assume contaminant must be remediated in rock matrix as well as in rock fractures.
B. Incorporate appropriate estimates of plume longevity, based on sound characterization, into remedial action plans.
C. Develop and embrace realistic regulatory frameworks for setting remedial objectives and formalizing the transition from active remediation to long-term monitoring.
D. Design monitoring programs based on sound characterization and realistic remedial objectives, ensuring they are dynamic and informed by the remediation process itself.
E. Include natural attenuation in all remediation designs, ensuring designs are based on a sound conceptual site model and realistic performance expectations (e.g., over time frames that may range from decades to centuries).
F. Produce detailed, publicly accessible, research-level documentation regarding the application of different remediation technologies in a variety of fractured rock settings. This should be practitioner driven and government facilitated. Analysis of long-term performance of alternate contaminant remediation and transport control would benefit practice.
G. Increase the degree and efficiency of monitoring at fractured rock sites, particularly during and following active remediation. Incorporate resulting data into feasibility studies at other sites.
H. Develop better information communication and transfer mechanisms within and between agencies responsible for addressing fractured rock issues. Shorter-term activities requiring fewer resources include developing sets of common attributes to incorporate into individual agency databases that allow for easier data mining. Longer-term activities requiring greater resources include creating a centralized data bank of fractured rock-related data and information.

Remediation in some fractured rock settings can be a decades-to-centuries process. Because monitoring over the duration of remedial activity is necessary, monitoring systems need to be not only robust and cost-effective, but also replaceable. Furthermore, advances in sensor technologies and reduction in their costs have created opportunities to increase the frequency and volume of data collection. This, in turn, creates opportunities for increased data feedback and refinement of site models and remediation plans. Thoughtful data analysis can suggest additional data collection

[1] See clu-in.org.

needs and drive field observations through a quantitative decision-analysis process. Large data volumes, however, can complicate data management and informative feedback.

Recommendation 9. Incorporate long-term performance into monitoring system design.

Engineering practitioners and regulators should expect monitoring systems to accommodate long-term performance and be

A. Durable and able to accommodate expected operation and maintenance;
B. Inclusive of meaningful sampling frequencies to monitor trends;
C. Designed to accommodate feedback so that monitoring strategies can be appropriately refined in response to new trends or findings;
D. Designed to accommodate long-term variations in climate, water levels, temperatures, and other site conditions expected over the remediation period;
E. Designed to require the minimal amounts of analytes to quantify performance effectiveness;
F. Designed to be cognizant of the implications of discrete pathways, rock matrix contaminant storage, and issues of geologic heterogeneity and anisotropy when point source concentration measurements are used; and
G. Capable of data storage and management such that data can be accessed in meaningful ways.

Research on automation of data collection, archiving, and retrieval and on the triggering of alarms when data values surpass tolerance levels is recommended. Research is also recommended on alternative monitoring approaches, in particular those for remote monitoring of field-assessable parameters that trigger additional sampling and analysis when needed. Best-practice protocols need to be established and communicated to future practitioners responsible for monitoring systems that may have been put in place decades or centuries earlier.

THE OBSERVATIONAL APPROACH

Adaptive and observational approaches to characterization of fractured rock sites are more effective than prescriptive, linear approaches. Site and contaminant characterization, remediation, and monitoring are most effective when data feedback allows for informed modifications in approach. Adaptive or observational methods recognize the value of information gathered once an engineering process is under way and formally integrate this information into engineering decision making. Guidelines for and examples of use of observational and adaptive approaches at fractured rock sites would facilitate greater use of these methods in practice. Specific steps in the process are available in the literature and discussed in Chapter 7.

Recommendation 10. Use observational methods and adaptive approaches to inform engineering decisions made for fractured rock sites.

Regular adjustments to monitoring plans and to fractured rock site engineering more generally, including modifications to performance criteria and data processing approaches, should be made as appropriate in response to new information. Engineering decisions should be informed by

A. Risk-based and performance-based criteria, as appropriate, in prioritizing the components of contaminated site management;
B. Monitoring and remediation approaches and criteria that test model assumptions concerning transport pathways and chemical and biological remediation processes; and
C. Systems analysis approaches to integrate the value of information and support decisions made under uncertainty.

FINAL THOUGHTS

Better engineering, better use of resources, and improved outcomes will result if the use of oversimplified site conceptual models are avoided and realistic expectations regarding outcomes are adopted. The recommendations provided in this report are high level and intended to help the practitioner, researcher, and decision maker embrace a more interdisciplinary approach to engineering in the fractured rock environment. Although the fractured rock environment can be complex, it is subject to predictable laws of nature. This report describes how existing tools can be used to increase the accuracy and reliability of engineering design and management given those interacting forces of nature. With interdisciplinary and adaptive approaches to fractured rock site characterization and management, it is possible to conceptualize and model the fractured rock environment with acceptable levels of uncertainty and reliability and to design systems to maximize remediation potential and monitor long-term performance. Advances in technology and science need to be incorporated into engineering practice as they develop.

References

Acar, Ö., H. Klammler, K. Hatfield, M.A. Newman, M.D. Annable, J. Cho, B.L. Parker, J.A. Cherry, P. Pehme, P. Quinn, and R. Kroeker. 2013. A stochastic model for estimating groundwater and contaminant discharges from fractured rock passive flux meter measurements. *Water Resources Research* 49(3):1277–1291.

Acuna, J.A., I. Ershaghi, and Y.C. Yortsos. 1995. Practical application of fractal pressure transient analysis in naturally fractured reservoirs. *SPE Formation Evaluation* 10(3):173–179.

Adamski, M., V. Kremesec, R. Kolhatkar, C. Pearson, and B. Rowan. 2005. LNAPL in fine-grained soils: Conceptualization of saturation, distribution, recovery, and their modeling. *Ground Water Monitoring & Remediation* 25(1):100–112.

Akin, S. 2001. Estimation of fracture relative permeabilities from unsteady state corefloods. *Journal of Petroleum Science and Engineering* 30(1):1–14.

Alaskar, M., M. Ames, R. Horne, K. Li, S. Connor, and Y. Cui. 2010. In-situ multifunction nanosensors for fractured reservoir characterization. *Transactions—Geothermal Resources Council* 34(2):1009–1019.

Alaskar M., M. Ames, R.N. Horne, K. Li, Y. Cui, C. Liu, and S. Connor. 2011. Smart nanosensors for in-situ temperature measurement in fractured geothermal reservoirs. *Transactions—Geothermal Resources Council* 35(2):1371–1381.

Alcolea, A., and P. Renard. 2010. Blocking Moving Windo algorithm: Conditioning multiple-point simulations to hydrogeological data. *Water Resources Research* 46(8):W08511. doi: 10.1029/2009 WR007943.

Allen-King, R.M., R.W. Gillham, and D.M. Mackay. 1996. Sorption of dissolved chlorinated solvents to aquifer material. In *Dense Chlorinated Solvents and Other DNAPLS in Groundwater: History, Behavior, and Remediation,* J.F. Pankow and J.A. Cherry, eds. Portland, OR: Waterloo Press. Pp. 233–266.

Alshuhail, A., and D.C. Lawton. 2007. *Time-lapse surface seismic monitoring of injected CO_2 at the Penn West CO2-EOR site, Violet Grove, Alberta.* Consortium for Research on Elastic Wave Exploration Seismology. Vol. 17, Ch 69. 19 pp. Available online at http://www.crewes.org/ForOurSponsors/ResearchReports/2007/2007-69.pdf (accessed August 4, 2015).

Amann, R.I., W. Ludwig, and K.-H. Schleifer. 1995. Phylogenetic identification and in situ detection of individual microbial cells without cultivation. *Microbiological Reviews* 59(1):143.

Ameli, P., J.E. Elkhoury, and R.L. Detwiler. 2013. High-resolution fracture aperture mapping using optical profilometry. *Water Resources Research* 49(10):7126–7132.

Ameli, P., J.E. Elchoury, J.P. Morris, and R.L. Detwiler. 2014. Fracture permeability alteration due to chemical and mechanical processes: A coupled high-resolution model. *Rock Mechanics and Rock Engineering* 47(5):1563–1573.

Anderson, M.P., and W.W. Woessner. 1992. *Applied groundwater modeling: Simulation of flow and advective transport.* San Diego, CA: Academic Press.

Andersson, P., J. Byegård, B. Dershowitz, T. Doe, J. Hermanson, P. Meier, E.-L. Tullborg, and A. Winberg, eds. 2002. *Final report of the TRUE Block Scale Project: 1. Characterization and model development.* Svensk Kärnbränslehantering AB (SKB) Technical Report TR-02-13. Available at http://www.skb.se/upload/publications/pdf/TR-02-13.pdf (accessed December 9, 2013).

Ang, A.H.-S., and W.H. Tang. 1975. *Probability concepts in engineering planning and design, Vol. 1, Basic principles.* New York: John Wiley.

Anifowose, F., J. Labadin, and A. Abdulraheem. 2013. Predicting petroleum reservoir properties from downhole sensor data using an ensemble model of neural networks. *ACM International Conference Proceeding Series*. 1st Workshop on Machine Learning for Sensory Data Analysis, MLSDA 2013; Dunedin; New Zealand; December 3, 2013. Pp. 27–34.

Annable, M.D., J.W. Jawitz, P.S.C. Rao, D.P. Dai, H. Kim, and A.L. Wood. 1998. Field evaluation of interfacial and partitioning tracers for characterization of effective NAPL-water contact areas. *Groundwater* 36(3):495–502.

Annable, M.D., K. Hatfield, J. Cho, H. Klammler, B.L. Parker, J.A. Cherry, and P.S.C. Rao. 2005. Field-scale evaluation of the passive flux meter for simultaneous measurement of groundwater and contaminant fluxes. *Environmental Science & Technology* 39(18):7194–7201.

Arnold, B.W., P. Brady, S. Altman, P. Vaughn, D. Nielson, J. Lee, F. Gibb, P. Mariner, K. Travis, W. Halsey, J. Beswick, and J. Tillman. 2012. Research, development, and demonstration roadmap for deep borehole disposal. FCRD-USED-2012-000269, SAND2012-8527P. Albuquerque, NM: Sandia National Laboratories.

Arnon, S., E. Adar, Z. Ronen, A. Yakirevich, and R. Nativ. 2005. Impact of microbial activity on the hydraulic properties of fractured chalk. *Journal of Contaminant Hydrology* 76(3–4):315–336.

Arnould, M. 2006. Discontinuity networks in mudstones: A geological approach: Implications for radioactive wastes isolation in deep geological formations in Belgium, France, Switzerland. *Bulletin of Engineering Geology and the Environment* 65(4):413–422. https://doi.org/10.1007/s10064-006-0060-3.

Austria, J.J.C., M.J. O'Sullivan, and J. Doherty. 2015. *Dual porosity models of a discharge test*. Proceedings World Geothermal Congress 2015. Melbourne, Australia, April 19–25.

Bäckblom, G., and Öhberg, A. 2002. *The observational method applied to engineering and construction of the access to the ONKALO facility*. Olkiluto, Finland: Posiva Oy. Working Report 2002-48.

Baghbanan, A., and L. Jing. 2007. Hydraulic properties of fractured rock masses with correlated fracture length and aperture. *International Journal of Rock Mechanics and Mining Sciences* 44(5):704–719.

Bahrani, N., and D.D. Tannant. 2011. Field-scale assessment of effective dilation angle and peak shear displacement for a footwall slab failure surface. *International Journal of Rock Mechanics and Mining Sciences* 48(4):565–579.

Barcelona, M.J., J.P. Gibb, J.A. Helfrich, and E.E. Garske. 1985. *Practical guide for ground-water sampling*. EPA/600/2-85/104 (NTIS PB86- 137304). 184 pp.

Barker, J.A. 1988. A generalized radial flow model for hydraulic tests in fractured rock. *Water Resources Research* 24(10):1796–1804.

Barnett, B., L.R. Townley, V. Post, R.E. Evans, R.J. Hunt, L. Peeters, S. Richardson, A.D. Werner, A. Knapton, and A. Boronkay. 2012. *Australian groundwater modeling guidelines*. Water Report Series No. 82, June. Canberra, Australia: National Water Commission.

Bartlett, R., S.H. Bottrell, K. Sinclair, S. Thornton, I.D. Fielding, and D. Hatfield. 2010. Lithological controls on biological activity and groundwater chemistry in quaternary sediments. *Hydrological Processes* 24(6):726–735.

Bartram, A.K., M.D.J. Lynch, J.C. Stearns, G. Moreno-Hagelsieb, and J.D. Neufeld. 2011. Generation of multimillion-sequence 16S rRNA gene libraries from complex microbial communities by assembling paired-end Illumina reads. *Applied and Environmental Microbiology* 77:3846–3852.

Bastona, D.P., and B.H. Kueper. 2009. Thermal conductive heating in fractured bedrock: Screening calculations to assess the effect of groundwater influx. *Advances in Water Resources* 32(2):231–238.

Baveye, P., and A. Valocchi. 1989. An evaluation of mathematical models of the transport of biologically reacting solutes in saturated soils and aquifers. *Water Resources Research* 25(6):1413–1421.

Bear, J. 1972. *Dynamics of fluids in porous media*. New York: Elsevier.

Bear, J., C.-F. Tsang, and G. de Marsily. 1993. *Flow and Contaminant Transport in Fractured Rock*. San Diego, CA: Academic Press.

Bear, J., C.-F. Tsang, and G. de Marsily. 2012. *Flow and Contaminant Transport in Fractured Rock*. New York: Elsevier Science.

Becker, M.W., and A.M. Shapiro. 2000. Tracer transport in fractured crystalline rock: Evidence of nondiffusive breakthrough tailing. *Water Resources Research 36*(7):1677–1686.

Becker, M.W., and G.P. Tsoflias. 2010. Comparing flux-averaged and resident concentration in a fractured bedrock using ground penetrating radar data. *Water Resources Research* 46:W09518. doi: 09510.01029/02009WR008260.

Becker, M.W., P.W. Reimus, and P. Vilks. 1999. transport and attenuation of carboxylate-modified latex microspheres in fractured rock laboratory and field tracer tests. *Groundwater* 37(3):387–395.

Berg, S.J., and W.A. Illman. 2013. Field study of subsurface heterogeneity with steady-state hydraulic tomography. *Groundwater* 51(1):29–40.

Berkowitz, B. 2002. Characterizing flow and transport in fractured geological media: A review. *Advances in Water Resources* 25(8–12):861–884.

Berkowitz, B., and H. Scher. 1995. On characterization of anomalous-dispersion in porous and fractured media. *Water Resources Research* 31(6):1461–1466.

Berryman, J. G., and H.F. Wang. 1995. The elastic coefficients of double-porosity models for fluid transport in jointed rock. *Journal of Geophysical Research: Solid Earth* 100(B12):24611–24627.

Beugelsdijk, L.J.L., C.J. de Pater, and K. Sato. 2000. *Experimental hydraulic fracture propagation in a multi-fractured medium.* Society of Petroleum Engineers. doi: 10.2118/59419-MS.

Bibby, R. 1981. Mass-transport of solutes in dual-porosity media. *Water Resources Research* 17(4):1075–1081.

Bijeljic, B., S. Rubin, H. Scher, and B. Berkowitz. 2011. Non-Fickian transport in porous media with bimodal structural heterogeneity. *Journal of Contaminant Hydrology* 120–121:213–221.

Binley, A., G. Cassiani, R. Middleton, and P. Winship. 2002. Vadose zone flow model parameterisation using cross-borehole radar and resistivity imaging. *Journal of Hydrology* 267:147–159.

Birkholzer, J.T., and Y.W. Tsang. 2000. Modeling the thermal-hydrologic processes in a large-scale underground heater test in partially saturated fractured tuff. *Water Resources Research* 36(6):1431–1447.

Birkholzer, J., H. Rubin, H. Daniels, and G. Rouve. 1993. Contaminant advection and spreading in a fractured permeable formation. Part 1. Parametric evaluation and analytical solution. *Journal of Hydrology* 144(1–4):1–33.

Bisdom, K., G. Bertotti, and H.M. Nick. 2015. Outcrop-based geomechanical fracture aperture and flow modeling: The importance of shear on flow. *Search and Discovery* (41606).

Black, J., M. Brightman, and D. Holmes. 1987. *Crosshole investigations: Hydrogeological results and interpretations*. Technical Report NTB 97-37. Baden, Switzerland: Nagra (National Cooperative for the Disposal of Radioactive Waste).

Blessent, D., R. Therrien, and K. MacQuarrie. 2009. Coupling geological and numerical models to simulate groundwater flow and contaminant transport in fractured media. *Computers and Geosciences* 35(9):1897–1906.

Blomqvist, R., J. Kaija, P. Lampinen, M. Paananen, T. Ruskeeniemi, J. Korkealaakso, P. Pitkänen, J.-E. Ludvigson, J. Smellie, L. Koskinen, E. Floría, M.J. Turrero, G. Galarza, K. Jakobsson, M. Laaksoharju, J. Casanova, B. Grundfelt, and P. Hernan. 1998. *The Palmottu natural analogue project. Phase I: Hydrogeological evaluation of the site*. Final report—Phase I. European Commission Nuclear Science and Technology Series EUR 18202 EN. Luxembourg: Office for Official Publications of the European Communities. 98 pp.

Bodin, J., G. Porel, F. Delay, F. Ubertosi, S. Bernard, and J.-R. de Dreuzy. 2007. Simulation and analysis of solute transport in 2D fracture/pipe networks: The SOLFRAC program. *Journal of Contaminant Hydrology* 89(1–2):1–28.

Bonneau, F., G. Caumon, P. Renard, and J. Sausse. 2013. *Hybrid discrete fracture network simulation driven by statistics, tectonic history and geomechanics*. 2nd EAGE Workshop Naturally Fractured Reservoirs: Naturally Fractured Reservoirs in Real Life; Muscat; Oman; December 8, 2013, through December 11, 2013.

Bordas, J.M. 2005. Modeling groundwater flow and contaminant transport in fractured aquifers. Thesis, Air Force Institute of Technology, Air University. Available at http://oai.dtic.mil/oai/oai?verb=getRecord&metadataPrefix=html&identifier=ADA435202 (accessed March 31, 2014).

Bradley, P.M., P.J. Lacombe, T.E. Imbrigiotta, F.H. Chapelle, and D.J. Goode. 2009. Flowpath independent monitoring of reductive dechlorination potential in a fractured rock aquifer. *Ground Water Monitoring and Remediation* 29(4):46–55.

Bredehoeft, J. 2005. The conceptualization model problem—surprise. *Hydrogeology Journal* 13(1):37–46.

Brown, S.R. 1995. Simple mathematical-model of a rough fracture. *Journal of Geophysical Research: Solid Earth* 100(B4):5941–5952.

Brown, S.R., and C.H. Scholz. 1985. Broad bandwidth study of the topography of natural rock surfaces. *Journal of Geophysical Research: Solid Earth* 90(B14):12575–12582.

Brown, S.R., and C.H. Scholz. 1986. Closure of rock joints. *Journal of Geophysical Research: Solid Earth* 91(B5):4939–4948.

Brown, S., A. Caprihan, and R. Hardy. 1998. Experimental observation of fluid flow channels in a single fracture. *Journal of Geophysical Research: Solid Earth (1978–2012)* 103(B3):5125–5132.

Bruggeman, D.A.G. 1935. Berechnung verschiedener physikalishcer Kontanten von heterogenen Substanzen. *Annalen der physic (Leipzig)* 24:636–679.

Bruines, P. 2003. *Laminar ground water flow through stochastic channel networks in rock*. PhD thesis, École Polytechnique Fédérale de Lausanne. Available at http://infoscience.epfl.ch/record/33212/files/EPFL_TH2736.pdf (accessed March 31, 2014).

Bundschuh, J., and R. Al-Khoury. 2014. Geological CO_2 sequestration and compressed air energy storage—an introduction. In *Sustainable Energy Developments*, R. Al-Khoury and J. Bundschuh, eds. Boca Raton, FL: CRC Press. Pp. 1–10.

Buscheck, T.A., and J.J. Nitao. 1993. Repositiory-heat-driven Hypothermal flow at Yucca Mountain, Part II: Large-Scale in situ heater tests. *Nuclear Technology* 104(3):449–471.

Caine, J. S., J. P. Evans, and C. B. Forster. 1996. Fault zone architecture and permeability structure. *Geology* 24(11):1025–1028.

Cameron, D.A., and L.J. Durlofsky. 2014. Optimization and data assimilation for geological carbon storage. In *Sustainable Energy Developments*, R. Al-Khoury and J. Bundschuh, eds. Boca Raton, FL: CRC Press. Pp. 357–388.

Cardiff, M., X. Liu, P.K. Kitanidis, J. Parker, and U. Kim. 2010. Cost optimization of DNAPL source and plume remediation under uncertainty using a semi-analytical model. *Journal of Contaminant Hydrology* 113(1):25–43.

Carrera, J., X. Sánchez-Vila, I. Benet, A. Medina, G. Galarza, and J. Guimerà. 1998. On matrix diffusion: Formulations, solution methods and qualitative effects. *Hydrogeology Journal* 6(1):178–190.

Casey, A. 2008. Carbon cemetery. *Canadian Geographic* (Jan/Feb):68.

Casini, G., F. Lapponi, S. Jonoud, O.P. Wennberg, O. Karoussi, and A. Larsen. 2012. The effect of fracture-controlled cementation on fluid flow in carbonate reservoirs. *Search and Discovery* (120043).

Chadwick, A., R. Arts, C. Bernstone, F. May, S. Thibeau, and P. Zweigel. 2008. *Best practices for the storage of CO_2 in saline aquifers: observations and guidelines from the SACS and CO_2STORE projects*. Nottingham, UK: British Geological Survey. 277 pp.

Chamberlin, T.C. 1897. The method of multiple working hypotheses. *Journal of Geology* 5(8):837–848.

Chang, Y.J., P.E. Long, R. Geyer, A.D. Peacock, C.T. Resch, K. Sublette, S. Pfiffner, A. Smithgall, R.T. Anderson, H.A. Vrionis, J.R. Stephen, R. Dayvault, I. Ortiz-Bernad, D.R. Lovley, and D.C. White. 2005. Microbial incorporation of C-13-labeled acetate at the field scale: Detection of microbes responsible for reduction of U(VI). *Environmental Science & Technology* 39(23):9039–9048.

Chapelle, F.H., P.J. Lacombe, and P.M. Bradley. 2012. Estimated trichloroethene transformation rates due to naturally occurring biodegradation in a fractured rock aquifer. *Remediation Journal* 22(2):7–20.

Charbeneau, A., K.S. Novakowski, and N. Ross. 2006. The effect of a biofilm on solute diffusion in fractured porous media. *Journal of Contaminant Hydrology* 85(3–4):212–228.

Chartrand, M.M., P.L. Morrill, G. Lacrampe-Couloume, and B.S. Lollar. 2005. Stable isotope evidence for biodegradation of chlorinated ethenes at a fractured bedrock site. *Environmental Science & Technology* 39(13):4848–4856.

Chaudhuri, A., H. Rajaram, and H. Viswanathan. 2008. Alteration of fractures by precipitation and dissolution in gradient reaction environments: Computational results and stochastic analysis. *Water Resources Research* 44(10):W10410.

Chaudhuri, A., H. Rajaram, and H. Viswanathan. 2012. Fracture alteration by precipitation resulting from thermal gradients: Upscaled mean aperture-effective transmissivity relationship. *Water Resources Research* 48(1):W01601.

Chen, G., W.A. Illman, D.L. Thompson, V.V. Vesselinov, and S.P. Neuman. 2000. Geostatistical, type curve and inverse analyses of pneumatic injection tests in unsaturated fractured tuffs at the Apache Leap Research Site near Superior, Arizona. In *Dynamics of Flow and Transport in Fractured Rocks,* B. Faybishenko, P.A. Witherspoon, and S.M. Benson, eds. Washington, DC: American Geophysical Union. Pp. 73–98.

Chen, J., S. Hubbard, J. Peterson, K. Williams, M. Fienen, P. Jardine, and D. Watson. 2006. Development of a joint hydrogeophysical inversion approach and application to a contaminated fractured aquifer. *Water Resources Research* 42(6).

Chen, S., and G.D. Doolen. 1998. Lattice Boltzmann method for fluid flows. *Annual Review of Fluid Mechanics* 30(1):329–364.

Chen, Z., S.P. Narayan, Z. Yang, and S.S. Rahman. 2000. An experimental investigation of hydraulic behavior of fractures and joints in granitic rock. *International Journal of Rock Mechanics and Mining Sciences* 37(7):1061–1071.

Cherry, J.A., B.L. Parker, and C. Keller. 2007. A new depth-discrete multilevel monitoring approach for fractured rock. *Ground Water Monitoring & Remediation* 27(2):57–70.

Chiu, S.N., D. Stoyan, W.S. Kendal, and J. Mecke. 2013. *Stochastic geometry and its applications.* Chichester, UK: John Wiley & Sons.

Cho, H.J., R.J. Fiacco, and M. Daly. 2002. Soil vapor extraction and chemical oxidation to remediate chlorinated solvents in fractured crystalline bedrock: Pilot study results and lessons learned. *Remediation* 12(2):35–50.

Cho, H.J., R.J. Fiacco, and M. Daly. 2008. Characterization of crystalline bedrock contaminated by dense nonaqueous liquid. *Ground Water Monitoring & Remediation* 28(2):49–59.

Christensen, N.I., and W.D. Mooney. 1995. Seismic velocity structure and composition of the continental crust—a global view. *Journal of Geophysical Research-Solid Earth* 100(B6):9761–9788.

Christian, J.T., C.C. Ladd, and G.B. Baecher. 1994. Reliability applied to slope stability analysis. *Journal of Geotechnical Engineering* 120(12):2180–2207.

Chuprakov, D.A., A.V. Akulich, E. Siebrits, and M. Thiercelin. 2011. Hydraulic-fracture propagation in a naturally fractured reservoir. *SPE Production & Operations* 26(1):88–97.

CL:AIRE. 2010. *TDP24: Application of Thermally Enhanced Soil Vapour Extraction (TESVE) to remediate the unsaturated zone at the Western Storage Area (WSA), Harwell.* Technology Demonstration Project Bulletin. London, UK: CL:AIRE.

Claria, J.J., G.H. Goldsztein, and J.C. Santamarina. 2012. AC diffusion: Transport in porous networks subjected to zero-time-average advective flow. *Transport in Porous Media* 93(1):51–61.

Clement, T.P., B.S. Hooker, and R.S. Skeen. 1996. Macroscopic models for predicting changes in saturated porous media properties caused by microbial growth. *Groundwater* 34(5):934–942.

Cohen, R.M., A.H. Vincent, J.W. Mercer, C.R. Faust, and C.P. Spalding. 1994. *Methods for monitoring pump-and-treat performance.* EPA/600/R-94/123. Ada, OK: Robert S. Kerr Environmental Research Laboratory, Office of Research and Development, U.S. Environmental Protection Agency.

Cohen, R.M., J.W. Mercer, R.M. Greenwald, and M. S. Beljin. 1997. *Design guidelines for conventional pump-and-treat systems.* Washington, DC: Office of Research and Development, Office of Solid Waste and Emergency Response, U.S. Environmental Protection Agency.

Cole, J., W.J. McElroy, J. Glasgow, G. Heron, J. Galligan, K. Parker, and E.F. Davis. 2008. In situ thermal treatment of trichloroethene at Marshall Space Flight Center. Abstract P-032 in *Remediation of Chlorinated and Recalcitrant Compounds—2008.* Proceedings of the Sixth International Conference on Remediation of Chlorinated and Recalcitrant Compounds (Monterey, CA: May 2008). Columbus, OH: Battelle Press.

Cornette, B.M., C. Telker, and A. de la Pena. 2012. *Refining discrete fracture networks with surface microseismic mechanism inversion and mechanism-driven event location*. Richardson, TX: Society of Petroleum Engineers. doi: 10.2118/151964-MS.

Crawford, B.R., D.R. Faulkner, and E.H. Rutter. 2008. Strength, porosity, and permeability development during hydrostatic and shear loading of synthetic quartz-clay fault gouge. *Journal of Geophysical Research: Solid Earth* 113(B3).

Crawford, J. 2010. *Radionuclide transport report for the safety assessment SR-Site*. Stockholm, Sweden: Swedish Nuclear Fuel and Waste Management Co.

Cvetkovic, V., and G. Dagan. 1994. Transport of kinetically sorbing solute by steady random velocity in heterogeneous porous formations. *Journal of Fluid Mechanics* 265:189–216.

Cvetkovic, V., J.O. Selroos, and H. Chen. 1999. Transport of reactive tracers in rock fractures. *Journal of Fluid Mechanics* 378:335–356.

Dagan, G., and V. Cvetkovic. 1996. Reactive transport and immiscible flow in geological media. 1. General theory. *Proceedings of the Royal Society of London Series A—Mathematical Physical and Engineering Sciences* 452(1945):285–301.

Dagan, G., and S.P. Neuman. 1997. *Subsurface flow and transport: A stochastic approach*. Paris, France: United Nations Educational, Scientific and Cultural Organization.

Dahan, O., R. Nativ, E.M. Adar, B. Berkowitz, and N. Weisbrod. 2000. On fracture structure and preferential flow in unsaturated chalk. *Groundwater* 38(3):444–451.

Darlington, R., L. Lehmicke, R.G. Andrachek, and D.L. Freedman. 2008. Biotic and abiotic anaerobic transformations of trichloroethene and cis-1,2-dichloroethene in fractured sandstone. *Environmental Science and Technology* 42(12):4323–4330.

Day-Lewis, F.D., and J.W. Lane, Jr. 2004. Assessing the resolution-dependent utility of tomograms for geostatistics. *Geophysical Research Letters* 31:L07503. doi: 07510.01029/02004GL019617.

Day-Lewis, F.D., and K. Singha. 2008. Geoelectrical inference of mass transfer parameters: Theoretical basis and numerical experiments. *Water Resources Research* 44:W05201. doi: 10.1029/2007 WR006750.

Day-Lewis, F.D., J.W. Lane, Jr., J.M. Harris, and S.M. Gorelick. 2003. Time-lapse imaging of saline-tracer transport in fractured rock using difference-attenuation tomography. *Water Resources Research* 39(10):1290. doi: 1210.1029/2002WR0011722.

Day-Lewis, F.D., J.W. Lane, Jr., and S.M. Gorelick. 2004. Combined interpretation of radar, hydraulic and tracer data from a fractured-rock aquifer near Mirror Lake, New Hampshire, USA. *Hydrogeology Journal* 14(1–2):1–14. doi: 10.1007/s10040-004-0372-y.

Day-Lewis, F.D., K. Singha, and A.M. Binley. 2005. Applying petrophysical models to radar traveltime and electrical-resistivity tomograms: Resolution-dependent limitations. *Journal of Geophysical Research* 110(B8):B08206. doi: 08210.01029/02004JB003569.

Day-Lewis, F.D., C.D. Johnson, F.L. Paillet, and K.J. Halford. 2011. A computer program for flow-log analysis of single holes (FLASH). *Groundwater* 49(6):926–931.

Dershowitz, W.S. 1985. *Rock joint systems*. Ph.D. thesis, Massachusetts Institute of Technology, Cambridge.

Dershowitz, W.S. 2006. *Hybrid discrete fracture network and equivalent continuum model for shaft sinking*. Golden Rocks, 2006: The 41st U.S. Symposium on Rock Mechanics (USRMS), June 17–21, Golden, Colorado. Available at https://www.onepetro.org/conference-paper/ARMA-06-1029 (accessed March 31, 2014).

Dershowitz, W.S., and H.H. Einstein. 1990. Tensile and shear fracturing in predominantly compressive stress fields—a review. *Engineering Geology* 29:149–171.

Dershowitz W., P. La Pointe, and T. Cladouhos. 1998. Derivation of fracture spatial pattern parameters from borehole data. *International Journal of Rock Mechanics & Mining Sciences* 35:508.

Dershowitz, W., J. Hermanson, S. Follin, and M. Mauldon. 2000. Fracture intensity measures in 1-D, 2-D, and 3-D at Aspo, Sweden. In *Proceedings of Pacific Rocks.* Houston, TX: American Rock Mechanics Association.

Dershowitz, W., A. Winberg, J. Hermanson, J. Byegård, E. Tullborg, P. Andersson, and M. Mazurek. 2003. *Äspö Hard Rock Laboratory: Äspö task force on modelling of groundwater flow and transport of solutes.* Task 6c. A semi-synthetic model of block scale conductive structures at the Äspö HRL. International Progress Report, IPR-03-13. Stockholm, Sweden: Swedish Nuclear Fuel and Waste Management Company.

Dershowitz, W.S., P.R. La Pointe, and T.W. Doe. 2004. Advances in discrete fracture network modeling. Available at https://www.clu-in.org/products/siteprof/2004fracrockconf/cdr_pdfs/indexed/group1/882.pdf (accessed March 31, 2014).

Dietrich, P., R. Helmig, M. Sauter, H. Hötzl, J. Köngeter, and G. Teutsch, eds. 2005. *Flow and transport in fractured porous media.* Berlin, Germany: Springer.

Dijk, P.E., and B. Berkowitz. 1999. Three-dimensional flow measurements in rock fractures. *Water Resources Research* 35(12):3955–3959.

Dijk, P.E., B. Berkowitz, and P. Bendel. 1999. Investigation of flow in water-saturated rock fractures using nuclear magnetic resonance imaging (NMRI). *Water Resources Research* 35(2):347–360.

Dobson, P.F., T.J. Kneafsey, E.L. Sonnenthal, N. Spycher, and J.A. Apps. 2003. Experimental and numerical simulation of dissolution and precipitation: implications for fracture sealing at Yucca Mountain, Nevada. *Journal of Contaminant Hydrology* 62(3):459–476.

Doe, T.W. 1991. *Fractional dimension analysis of constant-pressure well tests.* Presented at SPE Annual Technical Conference and Exhibition, October 6–9, Dallas, Texas.

Doherty, J.E., R.J. Hunt, and M.J. Tonkin. 2010. *Approaches to highly parameterized inversion: A guide to using PEST for model parameter and predictive-uncertainty analysis.* U.S. Geological Survey Scientific Investigations Report 2010–5211, 71 pp. Available at http://pubs.usgs.gov/sir/2010/5211/pdf/sir20105211_hunt_083111.pdf (accessed March 31, 2014).

Dorn, C., N. Linde, T.L. Borgne, O. Bour, and L. Baron. 2011. Single-hole GPR reflection imaging of solute transport in a granitic aquifer. *Geophysical Research Letters* 38:L08401. doi: 08410.01029/020 11GL047152.

Dradjat, A.S., X. Hu, and R. Primasari. 2012. Application of pre-stack seismic anisotropy for fracture detection, in Oseil Field Carbonate Reservoir, Seram Island, Eastern Indonesia. *Search and Discovery* (20149).

Dultz, S., A.V. Simonyan, J. Pastrana, H. Behrens, M. Plötze, and T. Rath. 2013. Implications of pore space characteristics on diffusive transport in basalts and granites. *Environmental Earth Sciences* 69(3):969–985.

Durham, W., and B. Bonner. 1993. PEAK: A new kind of surface microscope. *International Journal of Rock Mechanics and Mining Sciences & Geomechanics Abstracts* 30(7):699–702.

Durlofsky, L.J. 1992. Representation of grid block permeability in coarse scale models of randomly heterogeneous porous media. *Water Resources Research* 28(7):1791–1800.

Durlofsky, L.J. 2002. Upscaling of geological models for reservoir simulation: Issues and approaches. *Computational Geosciences* 6:1–4.

Dusseault, M.B. 2011. Geomechanical challenges in petroleum reservoir exploitation. *KSCE Journal of Civil Engineering* 15(4):669–678.

Dverstorp, B., and J. Andersson. 1989. Application of the discrete fracture network concept with field data—possibilities of model calibration and validation. *Water Resources Research* 25(3):540–550.

Dverstorp, B., J. Andersson, and W. Nordqvist. 1992. Discrete fracture network interpretation of field tracer migration in sparsely fractured rock. *Water Resources Research* 28(9):2327–2343.

Edwards, K.J., W. Bach, and T.M. McCollom. 2005. Geomicrobiology in oceanography: Microbe-mineral interactions at and below the seafloor. *Trends in Microbiology* 13(9):449–456.

Ehrlich, H.L. 1999. Microbes as geologic agents: their role in mineral formation. *Geomicrobiology Journal* 16(2):135–153.

Eiken, O., P. Ringrose, C. Hermanrud, B. Nazarian, T.A. Torp, and L. Høier. 2011. Lessons learned from 14 years of CCS operations: Sleipner, In Salah, and Snøhvit. *Energy Procedia* 4:5541–5548.

Einstein, H.H. 1988. *Landslide risk assessment*. Keynote Lecture. Proceedings 5th International Symposium on Landslides. Lausanne, Switzerland.

Einstein, H.H. 1995. Uncertainty and expert opinion in geologic hazards. *The Earth, Engineers, and Education: A symposium in honor of Robert V. Whitman*. Cambridge, Department of Civil and Environmental Engineering, Massachusetts Institute of Technology.

Einstein, H.H. 2002. *Risk assessment and management in geotechnical engineering*. Keynote Lecture. Proceedings 8th Portuguese Congress for Geotechnique, Lisbon, Portugal. Pp. 2237–2262.

Einstein, H.H., and G.B. Baecher. 1982. Probabilistic and statistical methods in engineering geology, Part I. Problem statement and introduction to solution. *Rock Mechanics* (Suppl. 12):47–61.

Einstein, H.H., D.A. Labreche, M.J. Markow, and G.B. Baecher. 1978. Decision analysis applied to rock tunnel exploration. *Engineering Geology* 12:143–161.

Eker, E., and S. Akin. 2006. Lattice Boltzmann simulation of fluid flow in synthetic fractures. *Transport in Porous Media* 65(3):363–384.

Elkhoury, J.E., P. Ameli, and R.L. Detwiler. 2013. Dissolution and deformation in fractured carbonates caused by flow of CO_2-rich brine under reservoir conditions. *International Journal of Greenhouse Gas Control: Supplement 1* 16:S203–S215.

Ellis, B.R., J.P. Fitts, G.S. Bromhal, D.L. McIntyre, R. Tappero, and C.A. Peters. 2013. Dissolution-driven permeability reduction of a fractured carbonate caprock. *Environmental Engineering Science* 30(4):187–193.

Emmanuel, S., and B. Berkowitz. 2005. Mixing-induced precipitation and porosity evolution in porous media. *Advances in Water Resources* 28(4):337–344.

EPA (U.S. Environmental Protection Agency). 1996. *Pump-and-Treat Ground-Water Remediation: A Guide for Decision Makers*. EPA/625/R-95/005. Cincinnati, OH: Office of Research and Development, National Risk Management Research Laboratory, Center for Environmental Research Information, U.S. Environmental Protection Agency.

EPA. 1999a. *A guide to preparing superfund proposed plans, records of decisions, and other remedy selection decision documents*. Office of Solid Waste and Emergency Response, EPA 540-R-98-031. 182 pp.

EPA. 1999b. *Use of monitored natural attenuation at Superfund, RCRA corrective action, and underground storage tank sites*. Available at http://www.epa.gov/swerust1/directiv/d9200417.pdf (accessed January 16, 2014).

EPA. 2001. *The state-of-the-practice of characterization and remediation of contaminated ground water at fractured rock sites*. EPA 542-R-01-010. Washington, DC: Office of Solid Waste and Emergency Response, U.S. Environmental Protection Agency. Available at http://purl.access.gpo. gov/GPO/LPS32622 (accessed January 16, 2014).

EPA. 2002. *Elements for effective management of operating pump and treat systems*. EPA 542-R-02-009. Washington, DC: Office of Solid Waste and Emergency Response, U.S. Environmental Protection Agency. Available at http://purl.access.gpo.gov/GPO/LPS29738 (accessed January 16, 2014).

EPA. 2004. *Cleaning up the nation's waste sites markets and technology trends*. Washington, DC: Office of Solid Waste and Emergency Response, U.S. Environmental Protection Agency. Available at http://purl.access.gpo.gov/GPO/LPS112739 (accessed January 16, 2014).

EPA. 2008. *A systematic approach for evaluation of capture zones at pump and treat systems*. EPA 600/R-08/003. Cincinnati, OH: Office of Research and Development, National Risk Management Research Laboratory, Ground Water and Ecosystems Restoration Division, U.S. Environmental Protection Agency. Available at http://purl.access.gpo.gov/GPO/LPS118015 (accessed January 16, 2014).

EPA. 2011. Groundwater road map: Recommended process for restoring contaminated groundwater at superfund sites. OSWER 9283.1-34. Available at http://www.epa.gov/superfund/health/conmedia/gwdocs/pdfs/gwroadmapfinal.pdf (accessed January 16, 2014).

EPA. 2012. *Superfund proposed plan*. Cayuga County Groundwater Contamination Superfund Site, Cayuga County. Available at http://www.epa.gov/region02/superfund/npl/cayuga/pdf/Cayuga_prap.pdf (accessed January 17, 2014).

EPA. 2014a. *Contaminated site clean-up information*. Available at http://clu-in.org/contaminantfocus/default.focus/sec/Fractured_Rock/cat/Overview (accessed January 20, 2014).

EPA. 2014b. *Groundwater remedy completion strategy: Moving forward with the end in mind*. OSWER Directive 9202.2-1-44. Available at http://www.epa.gov/superfund/health/conmedia/gwdocs/pdfs/EPA_Groundwater_Remedy_Completion.pdf (accessed January 20, 2014).

Evans, D.D., T.J. Nicholson, and T.C. Rasmussen, eds. 1987. *Flow and transport through unsaturated fractured rock*. Geophysical Monograph Series 42. Washington, DC: American Geophysical Union.

Evans, K.F., T. Kohl, L. Rybach, and R. J. Hopkirk. 1992. The effects of fracture normal compliance on the long-term circulation behavior of a hot dry rock reservoir: A parameter study using the new fully-coupled code fracture. In *Geothermal Resources Council Transactions*, G. Mata, ed. Pp. 449–456.

Farahani, M.A., and T. Gogolla. 1999. Spontaneous Raman scattering in optical fibers with modulated probe light for distributed temperature Raman remote sensing. *Journal of Lightwave Technology* 17(8):1379–1391.

Faybinshenko, B., P.A., Witherspoon, and S.M. Benson, eds. 2000. *Dynamics of fluids in fractured rock*. Geophysical Monograph Series 122. Washington, DC: American Geophysical Union.

Finsterle, S. 2004. Multiphase inverse modeling. *Vadose Zone Journal* 3(3):747–762.

Fisher, K., and N. Warpinski. 2011. Hydraulic fracture-height growth: Real data. *Society of Petroleum Engineers Production & Operations* 27(01):8–19.

Follin, S., L. Hartley, I. Rhén, P. Jackson, S. Joyce, D. Roberts, and B. Swift. 2014. A methodology to constrain the parameters of a hydrogeological discrete fracture network model for sparsely fractured crystalline rock, exemplified by data from the proposed high-level nuclear waste repository site at Forsmark, Sweden. *Hydrogeology Journal* 22(2):313–331.

Fossum, P., and R. Horne. 1982. Interpretation of tracer return profiles at Wairakei Geothermal Field using fracture analysis. *Transactions Geothermal Resource Council* 6:261–264.

Fourar, M., S. Bories, R. Lenormand, and P. Persoff. 1993. Two-phase flow in smooth and rough fractures: Measurement and correlation by porous-medium and pipe flow models. *Water Resources Research* 29(11):3699–3708.

Fox, A., P. La Pointe, and C. Enachescu. 2009. *Propagation of uncertainty from geological Discrete Fracture Networks (DFNs) to downstream models, upscaling of fractured carbonate reservoirs: Best practices and workflows*. Presentation at Society of Petroleum Engineers (SPE) Advanced Technology Workshop, Tunisia.

Frampton, A., and V. Cvetkovic. 2011. Numerical and analytical modeling of advective travel times in realistic three-dimensional fracture networks. *Water Resources Research* 47(2):W02506. doi: 10.1029/2010WR009290.

Fransson, Å. 2002. Nonparametric method for transmissivity distributions along boreholes. *Groundwater* 40(2):201–204.

Fredd, C.N., and H.S. Fogler. 1998. Influence of transport and reaction on wormhole formation in porous media. *AlChE Journal* 44(9):1933–1949.

Fredd, C.N., and H.S. Fogler. 1999. Optimum conditions for wormhole formation in carbonate porous media: Influence of transport and reaction. *SPE Journal* 4(03):196–205.

Fredrickson, J.K., J. P. McKinley, B.N. Bjornstad, P.E. Long, D.B. Ringelberg, D.C. White, L.R. Krumholz, J. M. Suflita, F. S. Colwell, and R.M. Lehman. 1997. Pore-size constraints on the activity and survival of subsurface bacteria in a late Cretaceous shale-sandstone sequence, northwestern New Mexico. *Geomicrobiology Journal* 14(3):183–202.

Freeze, R.A., and J.A. Cherry. 1979. *Groundwater*. Englewood Cliffs, NJ: Prentice-Hall Inc. 604 pp.

FRTR (Federal Remediation Technologies Roundtable). 2007. *Abstracts of remediation case studies*. Volume 11. 99 pp. Available at http://epa.gov/tio/download/frtr/volume11.pdf (accessed January 20, 2014).

Gallardo, L.A., and M.A. Meju. 2003. Characterization of heterogeneous near-surface materials by joint 2D inversion of DC resistivity and seismic data. *Geophysical Research Letters* 30(13).

Gascoyne, M. 2004. Hydrogeochemistry, groundwater ages and sources of salts in a granitic batholith on the Canadian Shield, southeastern Manitoba. *Applied Geochemistry* 19:519–560.

Gefell, M.J., K.L. Sperry, J.R.Y. Rawson, E. Kolodziej, and P.P. Mlodzinski. 2002. *Laboratory measurement of potassium permanganate consumption by sedimentary rock and field tracer test design*. Presented at the Third International Conference on Remediation of Chlorinated and Recalcitrant Compounds, May 20–23, Monterey, California.

Gefell, M.J., K.L. Sperry, J.R.Y. Rawson, E. Kolodziej, P. Mlodzinski, and C. Stanley. 2003. *Field-scale measurement of permanganate consumption rate and TCE oxidation in fractured bedrock—tracer test using permanganate and bromide*. Presented at National Ground Water Association Conference on Remediation—Site Closure and the Total Cost of Cleanup, November 13–14, New Orleans, Louisiana.

Gefell, M.J., K.L. Sperry, J.R.Y. Rawson, and E. Kolodziej. 2004. *Field-scale TCE oxidation in sedimentary bedrock: KMnO4 and Br- tracer test results and extended pilot design*. Presented at 2004 US EPA/NGWA Fractured Rock Conference: State of the Science and Measuring Success in Remediation, September 13–15, Portland, Maine.

Gefell, M.J., K.L. Sperry, D. Szuch, and E. Kolodziej. 2007. *Field-scale TCE oxidation in sedimentary bedrock: extended pilot application design and implementation*. Presented at 2007 US EPA/NGWA Fractured Rock Conference: State of the Science and Measuring Success in Remediation, September 24–26, 2007, Portland, Maine.

Gefell, M.J., D. Szuch, and E. Kolodziej. 2008. *Field-scale TCE oxidation in sedimentary bedrock: extended pilot application update*. Presented at the Sixth International Conference on Remediation of Chlorinated and Recalcitrant Compounds, May 19–22, 2008, Monterey, California.

Geller, J.T., H.Y. Holman, G. Su, M.E. Conrad, K. Pruess, and J. C. Hunter-Cevera. 2000. Flow dynamics and potential for biodegradation of organic contaminants in fractured rock vadose zones. *Journal of Contaminant Hydrology* 43(1):63–90.

Glass, R.J., M.J. Nicholl, H. Rajaram, and T.R. Wood. 2003. Unsaturated flow through fracture networks: Evolution of liquid phase structure, dynamics, and the critical importance of fracture intersections. *Water Resources Research* 39(12). doi: 10.1029/2003WR002015.

Goldstein, G.H., and J.C. Santamarina. 2004. Solute transport during cyclic flow in saturated porous media. *Applied Physics Letters* 85(12):2432–2434.

Goldstein, K.J., A.R. Vitolins, D. Navon, B.L. Parker, S.W. Chapman, and G.A. Anderson. 2004. Characterization and pilot-scale studies for chemical oxidation remediation of fractured shale. *Remediation Journal* 14(4):19–37.

Goldstein, K.J., A.R. Vitolins, D. Navon, S.W. Chapman, B.L. Parker, T.A. Al. 2007. Full-scale permanganate remediation of chlorinated ethenes in fractured shale: Part 1—site characterization and design and implementation of full-scale remedy. 2007 U.S. EPA /NGWA Fractured Rock Conference: State of the Science and Measuring Success in Remediation, Portland, Maine, September 24–26, 2007. Available at http://clu-in.org/products/siteprof/2007fracrock/018Goldstein,K.pdf (accessed January 20, 2014).

Golfier, F., C. Zarcone, B. Bazin, R. Lenormand, D. Lasseux, and M. Quintard. 2002. On the ability of a Darch-scale model to capture wormhole formation during the dissolution of a porous medium. *Journal of Fluid Mechanics* 457(1):213–254.

Golubev, V.S., and A.A. Garibyants. 1971. *Heterogeneous processes of geochemical migration*. New York: Consultants Bureau.

Gorelick, S.M. 1993. *Groundwater contamination: Optimal capture and containment*. Boca Raton, FL: Lewis Publishers.

Golser, H., and W. Schubert. 2003. Application of numerical simulation at the tunnel site. In *Numerical Simulation in Tunnelling*, Gernot Beer, ed. Chapter 15.

Grasmueck, M. 1996. 3-D ground-penetrating radar applied to fracture imaging in gneiss. *Geophysics* 61(4):1050–1064.

Grégoire, C., and L. Halleux. 2002. Characterization of fractures by GPR in a mining environment. *First Break* 20:467–471.

Griffin, T.W., and K.W. Watson. 2002. A comparison of field techniques for confirming dense nonaqueous phase liquids. *Ground Water Monitoring & Remediation* 22(2):48–59.

Griffin, T.W., T.J. Phelps, F.S. Colwell, and P.E. Long. 1997. Methods for obtaining deep subsurface microbiological samples by drilling. In *The Microbiology of the Terrestrial Deep Subsurface,* P.S. Amy and D.L. Haldeman, eds. Boca Raton, FL: CRC Press. Pp. 23–44.

Gringarten, A.C. 1987. Type-curve analysis: What it can and cannot do. *Journal of Petroleum Technology* 39(1):11–13.

Grisak, G.E., and J.F. Pickens. 1980. Solute transport through fractured media: 1. The effect of matrix diffusion. *Water Resources Research* 16(4):719–730.

Grob, M., and M. van der Baan. 2011. Inferring in-situ stress changes by statistical analysis of microseismic event characteristics. *The Leading Edge* 30(11):1296–1301.

Guerin, T.F., M. Mondido, B. McClenn, and B. Peasley. 2001. Application of resazurin for estimating abundance of contaminant-degrading micro-organisms. *Letters in Applied Microbiology* 32(5):340–345.

Habana, M.D. 2002. *Relative permeability of fractured rock*. Idaho Falls, ID: U.S. Department of Energy Idaho Operations Office. Available at http://www.osti.gov/servlets/purl/801752-XqwRc7/native (accessed January 20, 2014).

Hadley, P.W., and C. Newell. 2014. The new potential for understanding groundwater contaminant transport. *Journal of Groundwater* 52(2):174–186.

Haggerty, R., M.H. Schroth, and J.D. Istok. 1998. Simplified method of "push-pull" test data analysis for determining in situ reaction rate coefficients. *Groundwater* 36(2):314–324.

Haggerty, R., A. Argerich, and E. Martí. 2008. Development of a "smart" tracer for the assessment of microbiological activity and sediment-water interaction in natural waters: The resazurin-resorufin system. *Water Resources Research* 44(4).

Han, S.G., and C.F. Zhao. 2015. A new waterflooding characteristic curve for naturally fractured sandstone reservoirs. *Advances in Petroleum Exploration and Development* 9(1):9–12. http://dx.doi.org/10.3968/6405.

Haneberg, W.C., P.S. Mozley, J.C. Moore, and L.B. Goodwin. 1999. *Faults and subsurface fluid flow*. Geophysical Monograph Series 113. Washington, DC: American Geophysical Union.

Hansford, J., and Q. Fisher. 2009. The influence of fracture closure from petroleum production from naturally fractured reservoirs: A simulation modeling approach. *Search and Discovery* (40442).

Hartley, L., and S. Joyce. 2013. Approaches and algorithms for groundwater flow modeling in support of site investigations and safety assessment of the Forsmark site, Sweden. *Journal of Hydrology* 500:200–216.

Hartley, L., F. Hunter, P. Jackson, R. McCarthy, B. Gylling, and N. Marsic. 2006. *Regional hydrogeological simulations using CONNECTFLOW*. Preliminary site description Laxemar subarea—Version 1.2. SKB R-06-23, Svensk Kärnbränslehantering AB.

Hashin, Z., and S. Shtrikman. 1962. A variational approach to the theory of effective magnetic permeability of multiphase materials. *Journal of Applied Physics* 33:3125–3131.

Hatfield, K., M. Annable, J. Cho, P.S.C. Rao, and H. Klammler. 2004. A direct passive method for measuring water and contaminant fluxes in porous media. *Journal of Contaminant Hydrology* 75(3):155–181.

Heincke, B., T. Günther, E. Dalsegg, J.S. Rønning, G.V. Ganerød, and H. Elvebakk. 2010. Combined three-dimensional electric and seismic tomography study on the Åknes rockslide in western Norway. *Journal of Applied Geophysics* 70(4):292–306.

Helsen, J., R. Aravena, M. Zhang, O. Shouakar-Stash, and L. Burns. 2007. *Assessment of TCE Oxidation by KMnO4 using stable carbon and chlorine isotopes at a fractured bedrock site.* Fractured Rock Conference: State of the science and measuring success in remediation, September 24–26, 2007, Portland, ME. Available at http://www.clu-in.info/products/siteprof/2007fracrock/017Helsen,J.pdf (accessed January 20, 2014).

Herda, H.H.W., H.H. Einstein, and W.S. Dershowitz. 1991. Problems with representation of rock-fracture clusters. *Journal of Geotechnical Engineering* 117(11):1754–1771. https://doi.org/10.1061/(ASCE)0733-9410(1991)117:11(1754).

Hernqvist, L., S. Kvartsberg, A. Fransson, and G. Gustafson. 2012. A Swedish grouting design concept: Decision method for hard rock tunneling. *Geotechnical Special Publication* 228(228 GSP):816–825.

Heron, G., R.S. Baker, J.M. Bierschenk, and J.C. LaChance. 2008. Use of thermal conduction heating for the remediation of DNAPL in fractured bedrock. In *Remediation of Chlorinated and Recalcitrant Compounds: Proceedings of the Sixth International Conference* (Monterey, CA; May 2008). Columbus, OH: Battelle Press.

Hill, D.D., and B.E. Sleep. 2002. Effects of biofilm growth on flow and transport through a glass parallel plate fracture. *Journal of Contaminant Hydrology* 56(3-4):227–246.

Hinnell, A.C., T.P.A. Ferré, J.A. Vrugt, J.A. Huisman, S. Moysey, J. Rings, and M.B. Kowalsky. 2010. Improved extraction of hydrologic information from geophysical data through coupled hydrogeophysical inversion. *Water Resources Research* 46:W00D40. doi: 10.1029/2008WR007060.

Hjerne, C., R. Nordqvist, and J. Harrström. 2010. *Compilation and analyses of results from cross-hole tracer tests with conservative tracers*. Svensk Kärnbränslehantering AB (SKB) Report R-09-28. Available at http://www.skb.se/upload/publications/pdf/R-09-28.pdf (accessed December 10, 2013).

Hoek, E., and J. Bray. 1981. *Rock slope engineering*. London, UK: Institution of Mining and Metallurgy.

Hohnstock-Ashe, A.M., S.M. Plummer, R.M. Yager, P. Baveye, and E.L. Madsen. 2001. Further biogeochemical characterization of a trichloroethene-contaminated fractured dolomite aquifer: Electron source and microbial communities involved in reductive dechlorination. *Environmental Science and Technology* 35(22):4449–4456.

Hughes, T.R., M.J. Marton, A.R. Jones, C.J. Roberts, R. Stoughton, C.D. Armour, H.A. Bennett, E. Coffey, H. Dail, Y.D. He, M.J. Kidd, A.M. King, M.R. Meyer, D. Slade, P.Y. Lum, S.B. Stepaniants, D.D. Shoemaker, D. Gachotte, K. Chakraburtty, J. Simon, M. Bard, and S.H. Friend. 2000. Functional discovery via a compendium of expression profiles. *Cell* 102:109–126.

Huling, S.G., and J.W. Weaver. 1991. *Dense nonaqueous phase liquids*. Ada, OK: U.S. Environmental Protection Agency, Office of Research and Development, Office of Solid Waste and Emergency Response. Available at http://purl.access.gpo.gov/GPO/LPS49294 (accessed January 20, 2014).

Hurtig, E., S. Großwig, M. Jobmann, K. Kühn, and P. Marschall. 1994. Fibre-optic temperature measurements in shallow boreholes: Experimental application for fluid logging. *Geothermics* 23(4):355–364.

Ijiri, Y., H. Saegusa, A. Sawada, M. Ono, K. Watanabe, K. Karasaki, C. Doughty, M. Shimo, and K. Fumimura. 2009. Evaluation of uncertainties originating from the different modeling approaches applied to analyze regional groundwater flow in the Tono area of Japan. *Journal of Contaminant Hydrology* 103:168–181. Available at http://www.sciencedirect.com/science/article/pii/S0169772208001782 (accessed March 31, 2014).

Illman, W., and P.J.J. Alvarez. 2009. Performance assessment of bioremediation and natural attenuation. *Critical Reviews in Environmental Science and Technology* 39(4):209–270.

Illman, W.A. 2014. Hydraulic tomography offers improved imaging of heterogeneity in fractured rocks. *Groundwater* 52(5):659–684. doi: 10.1111/gwat.12119.

Illman, W.A., and S.P. Neuman. 2000. Type-curve interpretation of multi-rate single-hole pneumatic injection tests in unsaturated fractured rock. *Groundwater* 38(6):899–911.

Illman, W.A., and S.P. Neuman. 2001. Type-curve interpretation of a cross-hole pneumatic test in unsaturated fractured tuff. *Water Resources Research* 37(3):583–604.

Ilman, W.A., A.J. Craig, and X. Liu. 2008. Practical issues in imaging hydraulic conductivity through hydraulic tomography. *Groundwater* 46(1):120–132.

Illman, W.A., X. Liu, S. Takeuchi, T.C.J. Yeh, K. Ando, and H. Saegusa. 2009. Hydraulic tomography in fractured granite: Mizunami underground research site, Japan. *Water Resources Research* 45(1).

Ishibashi T., N. Watanabe, N. Hirano, A. Okamoto, and N. Tsuchiya. 2012. Upgrading of aperture model based on surface geometry of natural fracture for evaluating channeling flow. *Transactions—Geothermal Resources Council* 36(1):481–486.

Iwano, M., and H.H. Einstein. 1995. Laboratory experiments on geometric and hydromechanical characteristics of three different fractures in granodiorite. *8th ISRM Congress, 25–29 September, Tokyo, Japan*. ISRM-8CONGRESS-1995-151.

Jackson, C.P., A.R. Hoch, and S. Todman. 2000. Self-consistency of a heterogeneous continuum porous medium representation of a fractured medium. *Water Resources Research* 36:189–202.

Jenneman, G.E, M.J. McInerney, and R.M. Knapp. 1985. Microbial penetration through nutrient-saturated Berea sandstone. *Applied and Environmental Microbiology* 50(2):383–391.

Jiang, X.-W., L. Wan, X.-S. Wang, S.-H. Liang, and B.X. Hu. 2009. Estimation of fracture normal stiffness using a transmissivity-depth correlation. *International Journal of Rock Mechanics and Mining Sciences* 46(1):51–58.

Johnson, C.D., F.P. Haeni, J.W. Lane, and E.A. White. 2002. *Borehole-geophysical investigation of the University of Connecticut landfill*. Storrs, CT: U.S. Geological Survey. Water Resources Investigations Report 01-4033.

Johnson, P., P. Dahlen, J. Triplett Kingston, E. Foote, and S. Williams. 2009. State of the practice overview: Critical evaluation of state-of-the-art in situ thermal treatment technologies for DNAPL source zone treatment. ESTCP Project ER-0314. Available at https://www.clu-in.org/download/techfocus/thermal/Thermal-ER-0314-Overview.pdf (accessed January 20, 2014).

Johnston, J.J., M.A. Barlaz, and R.C. Borden. 1994. *Anaerobic biodegradation of hazardous organics in groundwater down gradient of a sanitary landfill*. Raleigh, NC: Water Resources Research Institute of the University of North Carolina.

Jones, G.A., J.-M. Kendall, I. Bastow, D.G. Raymer, and A. Wuestefeld. 2014. Characterization of fractures and faults: A multi-component passive microseismic study from the Ekofisk reservoir. *Geophysical Prospecting* 62(4):779–796.

Jurgawczynski, M. 2007. *Predicting absolute and relative permeabilities of carbonate rocks using image analysis and effective medium theory*. London, UK: University of London.

Kaldi, J.G., C.M. Gibson-Poole, and T.H.D. Payenberg. 2009. Geological inputs to selection and evaluation of CO_2 geosequestration sites. In *Carbon Dioxide Sequestration in Geological Media—State of the Science: AAPG Studies in Geology*, Volume 59, M. Grobe, J.C. Pashin, and R.L. Dodge, eds. Pp. 5–16.

Kazemi, H., and J.R. Gilman. 1993. Multiphase flow in fractured petroleum reservoirs. In *Flow and Contaminant Transport in Fractured Rock,* J. Bear, T. Chin-Fu, and G. de Marsily, eds. San Diego, CA: Academic Press.

Keller, K., and B.P. Bonner. 1985. Automatic, digital system for profiling rough surfaces. *Review of Scientific Instruments* 56(2):330.

Kelly, W.E. 1977. Geoelectric sounding for estimating aquifer hydraulic conductivity. *Groundwater* 15(6):420–425.

Kennel, J.R. 2008. *Advances in rock core VOC analyses for high resolution characterization of chlorinated solvent contamination in a dolostone aquifer*. Master's thesis, Department of Earth & Environmental Sciences, University of Waterloo, Waterloo, ON.

Keul, P.R., E. Mastin, J. Blanco, M. Maguérez, T. Bostick, and S. Knudsen. 2005. Using a fiber-optic seismic array for well monitoring. *The Leading Edge* 24(1):68–70.

Khilar, K.C., and H.S. Fogler. 1984. The existence of a critical salt concentration for particle release. *Journal of Colloid and Interface Science* 101:214–224.

Kia, S.F., H.S. Fogler, and M.G. Reed. 1987. Effect of pH on colloidally induced fines migration. *Journal of Colloid and Interface Science* 118:158–168.

Kilpatrick J.E., L. Eisner, and S. Williams-Stroud. 2010. Natural fracture characterization from microseismic source mechanisms: A comparison with FMI data. *SEG Technical Program Expanded Abstracts* 29(1):2110–2114.

Kim, J., and G.J. Moridis. 2015. Numerical analysis of fracture propagation during hydraulic fracturing operations in shale gas systems. *International Journal of Rock Mechanics and Mining Sciences* 76:127–137.

Kindred, J.S., and M.A. Celia. 1989. Contaminant transport and biodegradation: 2. Conceptual model and test simulations. *Water Resources Research* 25(6):1149–1159.

Kinner, N.E., R.W. Masters, and L.B. Fournier. 2005. *Fractured rock state of the science and measuring success in remediation.* Washington, DC: Office of Superfund Remediation and Technology Innovation, U.S. Environmental Protection Agency. Available at http://www.clu-in.org/download/remed/fractured_rock_summary.pdf (accessed January 17, 2014).

Klepikova, M.V., T. Le Borgne, O. Bour, and P. Davy. 2011. A methodology for using borehole temperature-depth profiles under ambient, single and cross-borehole pumping conditions to estimate fracture hydraulic properties. *Journal of Hydrology* 407(1):145–152.

Kleppe, J.H., and R.E. Olson. 1985. Desiccation cracking of soil barrier. In *Hydraulic Barrier in Soil and Rock*, A. Johnson, R. Frobel, N. Cavalli, and C. Pettersson, eds. West Conshohocken, PA: ASTM International. Pp. 263–275.

Klimczak, C., R.A. Schultz, R. Parashar, and D.M. Reeves. 2010. Cubic law with aperture-length correlation: Implications for network scale fluid flow. *Hydrogeology Journal* 18(4):851–862.

Klimentos, T., and C. McCann. 1990. Relationships among compressional wave attenuation, porosity, clay content, and permeability in sandstones. *Geophysics 55*(8):998–1014.

Knapp, R., M. Chiarappa, and W.B. Durham. 2000. An experimental exploration of the transport and capture of abiotic colloids in a single fracture. *Water Resources Research* 36:3139–3149.

Knothe Tate, M.L., U. Knothe, and P. Niederer. 1998. Experimental elucidation of mechanical load-induced fluid flow and its potential role in bone metabolism and functional adaptation. *American Journal of the Medical Sciences* 316(3):189–195.

Kowalsky, M.B., S. Finsterle, J. Peterson, S. Hubbard, Y. Rubin, E. Majer, A. Ward, and G. Gee. 2005. Estimation of field-scale soil hydraulic and dielectric parameters through joint inversion of GPR and hydrological data. *Water Resources Research* 41(11):W11425. doi: 10.1029/2005WR004237.

Kowalsky, M.B., J. Birkholzer, J. Peterson, S. Finsterle, S. Mukhopadhyay, and Y. Tsang. 2008. Sensitivity analysis for joint inversion of ground-penetrating radar and thermal-hydrological data from a large-scale underground heater test. *Nuclear Technology* 164(2):169–179.

Krumholz, L.R., J.P. McKinley, G.A. Ulrich, and J. M. Suflita. 1997. Confined subsurface microbial communities in Cretaceous rock. *Nature* 386(6620):64–66.

Kueper, B.H., and K.L. Davies. 2009. *Assessment and delineation of DNAPL source zones at hazardous waste sites*. EPA/600/R-09/119. Washington, DC: U.S. Environmental Protection Agency.

Kueper, B.H., and D.B. McWhorter. 1991. The behavior of dense, nonaqueous phase liquids in fractured clay and rock. *Groundwater* 29(5):716–728.

Kueper, B.H., D. Redman, R.C. Starr, S. Reitsma, and M. Mah. 1993. A field experiment to study the behavior of tetrachloroethylene below the water table—spatial distribution of residual and pooled DNAPL. *Groundwater* 31(5):756–766.

Lacasse, S., K. Hoeg, and F. Nadim. 2012. Risk assessment and mitigation in geo-practice. *Geotechnical Special Publication* 226(226 GSP):729–764.

Lacombe, P.J. 2000. *Hydrogeologic framework, water levels, and trichloroethylene contamination, Naval Air Warfare Center, West Trenton, New Jersey, 1993–1997.* U.S. Geological Survey Water-Resources Investigations Report 98-4167. P. 139.

Laine, J., M. Labady, A. Albornoz, and S. Yunes. 2008. Porosities and pore sizes in coralline calcium carbonate. *Materials Characterization* 59(10):1522–1525.

Landa, J.L., and D. Kumar. 2011. Joint inversion of 4D seismic and production data. *Proceedings—SPE Annual Technical Conference and Exhibition* 3:2615–2631.

Lane, Jr., J.W., F.P. Haeni, and W.M. Watson. 1995. Use of a square-array direct-current resistivity method to detect fractures in crystalline bedrock in New Hampshire. *Groundwater* 33(3):476–485.

Lane, Jr., J.W., F.P. Haeni, G. Placzek, and D.L. Wright. 1996. *Use of borehole-radar methods to detect a saline tracer in fractured crystalline bedrock at Mirror Lake, Grafton County, New Hampshire*. Paper presented at the Sixth International Conference on Ground-Penetrating Radar (GPR'96), September 30–October 3, 1996, Tohoku University Department of Geoscience and Technology, Sendai, Japan.

Lane, Jr., J.W., F.P. Haeni, and R. Versteeg. 1998. *Use of multi-offset borehole-radar reflection method in fractured crystalline bedrock at Mirror Lake, Grafton County, New Hampshire*. Paper presented at the Symposium on the Application of Geophysics to Engineering and Environmental Problems, Environmental and Engineering Geophysical Society, March 22–26, 1998, Chicago, Illinois.

Lane, Jr., J.W., F.D. Day-Lewis, J.M. Harris, F.P. Haeni, and S.M. Gorelick. 2000. *Attenuation-difference RADAR tomography: Results of a multiple-plan experiment at the U.S. Geological Survey Fractured Rock Research Site, Mirror Lake, New Hampshire*. Paper presented at GPR 2000—Proceedings of the Eighth International Conference on Ground Penetrating Radar, University of Queensland, Queensland, Australia.

Lebron, C.A., D. Phelan, G. Heron, J. LaChance, S.G. Neilsen, B.H. Kueper, D. Rodriguez, A. Wemp, D. Baston, P. LaCombe, and F.H. Chapelle. 2012. *Dense Non-Aqueous Phase Liquid (DNAPL) removal from fractured rock using Thermal Conductive Heating (TCH)*. ESTCP Project ER-200715, Contract Report CR-NAVFAC ESC-EV-1202. 427 pp. Alexandria, VA: Environmental Security Technology Certification Program.

Legchenko, A., J.M. Baltassat, A. Beauce, and J. Bernard. 2002. Nuclear magnetic resonance as a geophysical tool for hydrogeologists. *Journal of Applied Geophysics* 50(1):21–46.

Lehmen, R.M., F.F. Roberto, D. Earley, D.F. Bruhn, S.E. Brink, S.P. O'Connell, M.E. Delwiche, and F.S. Cowell. 2001. Attached and unattached bacterial communities in a 120-meter corehole in an acidic, crystalline rock aquifer. *Applied and Environmental Microbiology* 67(5):2095–2106.

Lenczewski, M., P. Jardine, L. McKay, and A. Layton. 2003. Natural attenuation of trichloroethylene in fractured shale bedrock. *Journal of Contaminant Hydrology* 64(3–4):151–168.

Lenormand, R., E. Touboul, and C. Zarcone. 1988. Numerical models and experiments on immiscible displacements in porous media. *Journal of Fluid Mechanics* 189(1):165–187.

Lian, P.Q., L.S. Cheng, and C.Y. Ma. 2012. The characteristics of relative permeability curves in naturally fractured carbonate reservoirs. *Journal of Canadian Petroleum Technology* 51(2):137–142.

Lima, G., B. Parker, and J. Meyer. 2012. Dechlorinating microorganisms in a sedimentary rock matrix contaminated with a mixture of VOCs. *Environmental Science and Technology* 46(11):5756–5763.

Linde, N., A. Binley, A. Tryggvason, L.B. Pedersen, and A. Revil. 2006. Improved hydrogeophysical characterization using joint inversion of cross-hole electrical resistance and ground-penetrating radar traveltime data. *Water Resources Research* 42:W12404. doi: 12410.11029/12006WR005131.

Lines, L.R., A.K. Schultz, and S. Treitel. 1988. Cooperative inversion of geophysical data. *Geophysics 53*(1):8–20.

Lipson, D.S., B.H. Kueper, and M.J. Gefell. 2005. Matrix diffusion-derived plume attenuation in fractured bedrock. *Groundwater* 43(1):30–39.

Liu, H.H., J. Houseworth, J. Rutqvist, L. Li, D. Asahina, F. Chen, and J. Birkholzer. 2012. *Report on modeling coupled processes in the near field of a clay repository*. Berkeley, CA: Lawrence Berkeley National Laboratory. FCRD-UFD-2012-000223. Available at http://energy.gov/sites/prod/files/2013/06/f1/FY12%20Report%20on%20Modeling%20Coupled%20Processes%20in%20the%20Near%20Field%20of%20a%20Clay%20Repository.pdf (accessed January 20, 2014).

Lojkasek-Lima, P., R. Aravena, B.L. Parker, and J.A. Cherry. 2012. Fingerprinting TCE in a bedrock aquifer using compound-specific isotope analysis. *Groundwater* 50(5):754–764.

Loman, N.J., C. Constantinidou, J.Z.M. Chan, M. Halachev, M. Sergeant, C.W. Penn, E.R. Robinson, and M.J. Pallen. 2012. High-throughput bacterial genome sequencing: An embarrassment of choice, a world of opportunity. *Nature Reviews Microbiology* 10:599–606.

Long, J.C.S. 1983. *Investigation of equivalent porous medium permeability in networks of discontinuous fractures*. Ph.D. dissertation, University of California, Berkeley.

Long, J.C.S., and P.A. Witherspoon. 1985. The relationship of the degree of interconnection to permeability in fracture networks. *Journal of Geophysical Research: Solid Earth and Planets* 90(4):3087–3098.

Long, J.C.S., J.S. Remer, C.R. Wilson, and P.A. Witherspoon. 1982. Porous media equivalents for networks of discontinuous fractures. *Water Resources Research* 18(3):645–658.

Longino, B.L., and B.H. Kueper. 1999. Nonwetting phase retention and mobilization in rock fractures. *Water Resources Research* 35(7):2085–2093.

Loucks, R.G., R.M. Reed, S.C. Ruppel, and D.M. Jarvie. 2009. Morphology, genesis, and distribution of nanometer-scale pores in siliceous mudstones of the Mississippian Barnett Shale. *Journal of Sedimentary Research* 79(11–12):848–861.

Lu, Z., and E.M. Kwicklis. 2012. Numerical evaluation of effective unsaturated hydraulic properties of fractured rocks using a stochastic continuum approach. *Vadose Zone Journal* 11(4). Available at http://www.ees.lanl.gov/staff/zhiming/Paper2012/uzeff_vzj.pdf (accessed March 31, 2014).

Macbeth, T.W., D.E. Cummings, S. Spring, L.M. Petzke, and K.S. Sorenson. 2004. Molecular characterization of a dechlorinating community resulting from in situ biostimulation in a trichloroethene-contaminated deep-fractured basalt aquifer and comparison to a derivative laboratory culture. *Applied and Environmental Microbiology* 70(12):7329–7341.

MacQuarrie, K.T.B, and K.U. Mayer. 2005. Reactive transport modeling in fracture rock: A state-of-the-science review. *Earth-Science Reviews* 72(3–4):189–227.

Mahmoudzadeh, B., L. Liu, L. Moreno, and I. Neretnieks. 2013. Solute transport in fractured rocks with stagnant water zone and rock matrix composed of different geological layers-model development and simulations. *Water Resources Research* 49(3):1709–1727.

Maloszewski, P. and A. Zuber. 1993. Tracer experiments in fractured rocks: Matrix diffusion and the validity of models. *Water Resources Research* 29(8):2723–2735.

Marinos, P., and E. Hoek. 2000. GSI—A geologically friendly tool for rock mass strength estimation. *Proceeding of the GeoEng2000 Conference*, Melbourne. Pp. 1422–1442.

Masciopinto, C. 2007. Biodegradation of wastewater nitrogen compounds in fractures: Laboratory tests and field observations. *Journal of Contaminant Hydrology* 92(3–4):230–254.

Matteo, C., P. Candido, R. Vera, and V. Francesca. 2012. Current and future nanotech applications in the oil industry. *American Journal of Applied Sciences* 9(6):784–793.

Matthäi, S.K., and H.M. Nick. 2009. Upscaling two-phase flow in naturally fractured reservoirs. *AAPG Bulletin* 93(11):1621–1632.

Mauldon, M. 1998. Estimating mean fracture trace length and density from observations in convex windows. *Rock Mechanics and Rock Engineering* 31(4):201–216. https://doi.org/10.1007/s006030050021.

Mauldon, M., and J.G. Mauldon. 1997. Fracture sampling on a cylinder: From scanlines to boreholes and tunnels. *Rock Mechanics and Rock Engineering* 30(3):129–144.

Mavko G., T. Mukerji, and J. Dvorkin. 2009. *The rock physics handbook: Tools for seismic analysis of porous media*. Cambridge, UK: Cambridge University Press.

Mazurek, M., A. Jakob, and P. Bossart. 2003. Solute transport in crystalline rocks at Äspö-I: Geological basis and model calibration. *Journal of Contaminant Hydrology* 61(1–4):157–174.

McCarthy, J.F., L.D. McKay, and D.D. Bruner. 2002. Influence of ionic strength and cation charge on transport of colloidal particles in fractured shale saprolite. *Environmental Science & Technology* 36(17):3735–3743.

McKay, L.D., J.A. Cherry, and R.W. Gillham. 1993. Field Experiments in a fractured clay till. 1. Hydraulic conductivity and fracture aperture. *Water Resources Research* 29(4):1149–1162.

Meacham, P.G., D.R. Anderson, E.J. Bonano, and M.G. Marietta. 2011. *Sandia National Laboratories performance assessment methodology for long-term environmental programs: The history of nuclear waste management*. SAND2011-8270. Albuquerque, NM: Sandia National Laboratories.

Min, K.B., J. Rutqvist, and C.F. Tsang. 2004. Stress-dependent permeability of fractured rock masses: a numerical study. *International Journal of Rock mechanics and Mining Sciences* 41(7):191–210.

Mitchell, J.K., and J.C. Santamarina. 2005. Biological considerations in geotechnical engineering. *Journal of Geotechnical and Geoenvironmental Engineering* 131(10):1222–1233.

Moran, M.A., B. Satinsky, S.M. Gifford, H. Luo, A. Rivers, L.-K. Chan, J. Meng, B.P. Durham, C. Shen, V.A. Varaljay, C.B. Smith, P.L. Yager, and B.M. Hopkinson. 2013. Sizing up metatranscriptomics. *ISME Journal* 7:237–243.

Moreno, L., and I. Neretnieks. 1993. Flow and nuclide transport in fractured media: the importance of the flow-wetted surface for radionuclide migration. *Journal of Contaminant Hydrology* 13:49–71.

Morgan, M.G., S.T. McCoy, J. Apt, M. Dworkin, P.S. Fischbeck, D. Gerard, K.A. Gregg, R.L. Gresham, C.R. Hagan, D.M. Kreis, R.R. Nordhaus, E.R. Pitlick, M. Pollak, J.S. Reiss, E.S. Rubin, K. Twaite, and E.J. Wilson. 2012. *Carbon capture and sequestration: Removing the legal and regulatory barriers*. New York: Resources for the Future Press.

Mosquera, M.J., T. Rivas, B. Prieto, and B. Silva. 2000. Microstructural changes in granitic rocks due to consolidation treatments: Their effects on moisture transport. *Proceedings of the 9th International Congress on Deterioration and Conservation of Stone* 601–607.

Moysey, S., and R.J. Knight. 2004. Modeling the field-scale relationship between dielectric constant and water content in heterogeneous systems. *Water Resources Research* 40:W03510. doi: 03510.01029/02003WR002589.

Moysey, S., K. Singha, and R. Knight. 2005. A framework for inferring field-scale rock physics relationships through numerical simulation. *Geophysical Research Letters* 32:L08304. doi: 08310.01029/02004GL022152.

Mundle, K., D.A. Reynolds, M.R. West, and B.H. Kueper. 2007. Concentration rebound following in situ chemical oxidation in fractured clay. *Groundwater* 45(6):692–702.

Narr, W., and J. Suppe. 1991. Joint spacing in sedimentary rocks. *Journal of Structural Geology* 13(9):1037–1048.

Nelson, R.A. 2001. *Geologic analysis of naturally fractured reservoirs*. Boston, MA: Gulf Professional Pub. Available at http://app.knovel.com/web/toc.v/cid:kpGANFRE01 (accessed January 20, 2014).

Neretnieks, I. 1980. Diffusion in the rock matrix: An important factor in radionuclide retardation. *Journal of Geophysical Research* 85(B8):4379–4397.

Neretnieks, I. 2002. A stochastic multi-channel model for solute transport—analysis of tracer tests in fractured rock. *Journal of Contaminant Hydrology* 55(3–4):175–211.

NETL (National Energy Technology Laboratory). 2012. *Best practices for carbon storage systems and well management activities*. Available at http://www.ourenergypolicy.org/wp-content/uploads/2012/06/BPM-Carbon-Storage-Systems-and-Well-Mgt.pdf (accessed January 20, 2014).

Neuman, S.P. 1987. Stochastic continuum representation of fractured rock permeability as an alternative to the REV and fracture network concepts. In *Rock Mechanics: Proceedings of the 28th US Symposium, Tucson, AZ*, I.W. Farmer, J.J.K. Daemen, C.S. Desai, C.E. Glass, and S.P. Neuman, eds. Rotterdam, The Netherlands: A.A. Balkema. Pp. 533–561.

Neuman, S.P. 1990. Universal scaling of hydraulic conductivities and dispersivities in geologic media. *Water Resources Research* 26(8):1749–1758.

Neuman, S.P. 2003. Maximum likelihood Bayesian averaging of alternative conceptual-mathematical models. *Stochastic Environmental Research and Risk Assessment* 17(5):291–305. doi: 10.1007/s00477-003-0151-7.

Neuman, S.P. 2005. Trends, prospects and challenges in quantifying flow and transport through fractured rocks. *Hydrogeology Journal* 13(1):124–147. doi: 10.1007/s10040-004-0397-2.

Neuman, S.P., and P.J. Wierenga. 2003. *A comprehensive strategy of hydrogeologic modeling and uncertainty analysis for nuclear facilities and sites.* NUREG/CR-6805. Washington, DC: U.S. Nuclear Regulatory Commission.

Neuman, S.P., L. Xue, M. Ye, and D. Lu. 2012. Bayesian analysis of data-worth considering model and parameter uncertainty. *Advances in Water Resources* 36:75–85.

Neuville, A., R. Toussaint, and J. Schmittbuhl. 2011. Hydraulic transmissivity and heat exchange efficiency of open fractures: a model based on lowpass filtered apertures. *Geophysical Journal International* 186(3):1064–1072.

Newell, C.J. 1995. *Light nonaqueous phase liquids.* Ada, OK: Office of Research and Development, Office of Solid Waste and Emergency Response, U.S. Environmental Protection Agency. Available at http://purl.access.gpo.gov/GPO/LPS49296 (accessed January 20, 2014).

Nicholl, M.J., R.J. Glass, and S.W. Wheatcraft. 1994. Gravity-driven infiltration instability in initially dry nonhorizontal fractures. *Water Resources Research* 30(9):2533–2546.

Nielsen, K.A. 2007. *Fractured aquifers: Formation evaluation by well testing.* Victoria, BC: Trafford.

Nimmer, R.E., J.L. Osiensky, A.M. Binley, K.F. Sprenke, and B.C. Williams. 2007. Electrical resistivity imaging of conductive plume dilution in fractured rock. *Hydrogeology Journal* 15(5):877–890.

Niva, B., O. Olsson, and P. Blumlung. 1988. *Grimsel test site: Radar crosshole tomography with application to migration of saline tracer through fracture zones.* TR88-31. Uppsala, Sweden: Swedish Geological Co.

Noiriel, C., B. Madé, and P. Gouze. 2007. Impact of coating development on the hydraulic and transport properties in argillaceous limestone fracture. *Water Resources Research* 43(9):W09406.

Nordqvist, A.W., Y.W. Tsang, C.F. Tsang, B. Dverstorp, and J. Andersson. 1992. A variable aperture fracture network model for flow and transport in fractured rocks. *Water Resources Research* 28(6):1703–1713.

Novakowski, K., P. Lapcevic, G. Bickerton, J. Voralek, L. Zanini, and C. Talbot. 2000. *The development of a conceptual model for contaminant transport in the dolostone underlying Smithville, Ontario.* Final report prepared for the Smithville Phase IV Remediation Program (available on CD). 175 pp.

NRC (National Research Council). 1994. *Alternatives for ground water cleanup.* Washington, DC: National Academy Press.

NRC. 1996. *Rock fractures and fluid flow: Contemporary understanding and applications.* Washington, DC: National Academy Press.

NRC. 1999. *Groundwater and soil cleanup: Improving management of persistent contaminants.* Washington, DC: National Academy Press.

NRC. 2000. *Natural attenuation for groundwater remediation.* Washington, DC: National Academy Press.

NRC. 2001. *Conceptual models of flow and transport in the fractured vadose zone.* Washington, DC: The National Academies Press.

NRC. 2003. *Environmental cleanup at Navy facilities: Adaptive site management.* Washington, DC: The National Academies Press.

NRC. 2007. *Assessment of the performance of engineered waste containment barriers.* Washington, DC: The National Academies Press.

NRC. 2013a. *Induced seismicity potential in energy technologies.* Washington, DC: The National Academies Press.

NRC. 2013b. *Alternatives for managing the nation's complex contaminated groundwater sites.* Washington, DC: The National Academies Press.

Odling, N.E., and J.E. Roden. 1997. Contaminant transport in fractured rocks with significant matrix permeability, using natural fracture geometries. *Journal of Contaminant Hydrology* 27(3–4):263–284.

Okada, H. 2003. *The Microtremor Survey Method.* Tulsa, OK: Society of Exploration Geophysicists with the cooperation of Society of Exploration Geophysicists of Japan and Australian Society of Exploration Geophysicists.

Olofsson, I., R. Christiansson, M. Holmberg, A. Carlsson, and D. Martin. 2014. *Application of the observational method in the Äspö expansion project*. Stockholm, Sweden: Swedish Nuclear Fuel and Waste Management Co. SKB R-13-44.

Olsson, O., L. Falk, O. Forslund, L. Lundmark, and E. Sandberg. 1992. Borehole radar applied to the characterization of hydraulically conductive fracture zones in crystalline rock. *Geophysical Prospecting* 40(2):109–142.

Ouenes, A., and L.J. Hartley. 2000. Integrated fractured reservoir modeling using both discrete and continuum approaches. *SPE Annual Technical Conference and Exhibition*. October 1–4, 2000. 62939. Dallas, TX: Society of Petroleum Engineers.

Paillet, F.L. 1998. Flow modeling and permeability estimation using borehole flow logs in heterogeneous fractured formations. *Water Resources Research* 34(5):997–1010.

Palcsu, L., E. Svingo, M. Mihaly, Z. Szanto, I. Futo, L. Rinyu, I. Horvath, G. Toth, and I. Foriz. 2007. Isotope studies of a groundwater-flow system in granite, Middle Hungary. In *Groudwater in Fractured Rocks: Selected papers from the Groundwater in Fractured Rocks International Conference, Prague, 2003*, J. Krasny and J.M. Sharp, eds. Boca Raton, FL: CRC Press.

Pampanelli, P.C.P., P.M. Silva, and M. Gattass. 2013. *A new volumetric fault attribute based on first order directional derivatives*. Thirteenth International Congress of the Brazilian Geophysical Society. Rio de Janeiro, Brazil, August 26–29, 2013. Available at http://sys2.sbgf.org.br/congresso/abstracts/trabalhos/sbgf_4240.pdf (accessed January 20, 2014).

Pang, T.W. 2010. *DNAPL remediation of fractured rock evaluated via numerical simulation*. Ph.D. thesis, College of Science and Engineering, The University of Edinburgh, Edinburgh, Scotland.

Park, J., G. Bolognini, D. Lee, P. Kim, P. Cho, F. D. Pasquale, and N. Park. 2006. Raman-based distributed temperature sensor with simplex coding and link optimization. *IEEE Photonics Technology Letters* 18(17):1879–1881.

Parker, B.L. 2007. Investigated contaminated sites on fractured rock using the DFN approach. *Proceedings of the 2007 U.S. EPA/NGWA Fractured Rock Conference, September 24–26*. 18 pp.

Parker, B.L., R.W. Gillham, and J.A. Cherry. 1994. Diffusive disappearance of immiscible-phase organic liquids in fractured geologic media. *Groundwater* 32(5):805–820.

Parker, B.L., D.B. McWhorter, and J.A. Cherry. 1997. Diffusive loss of non-aqueous phase organic solvents from idealized fracture networks in geologic media. *Groundwater* 35(6):1077–1088.

Parker, B.L., S.W. Chapman, and J.A. Cherry. 2010. Plume persistence in fractured sedimentary rock after source zone removal. *Groundwater* 48(6).

Patel, D., D. Nicholson, N. Huybrechts, and J. Maertens. 2007. The observational method in geotechnics. *Proceedings of the 14th ECSMGE* 2:365–370.

Payne, F.C., J.A. Quinnan, and S.T. Potter. 2008. *Remediation Hydraulics*. Boca Raton, FL: CRC Press.

Pearson, S.C., B.B. Johnson, N. Walter, R. Galloway, and S. Waldo. 2004. *Lessons learned from bedrock blast fracturing and bioremediation at a superfund landfill*. Fractured Rock Conference: State of the Science and Measuring Success in Remediation, September 13–15, Portland, ME.

Peck, R.B. 1969. Advantages and limitations of the observational method in applied soil mechanics. *Geotechnique* 19(2):171–187.

Pedersen, K. 1996. Investigations of subterranean bacteria in deep crystalline bedrock and their importance for the disposal of nuclear waste. *Canadian Journal of Microbiology* 42(4):382–391.

Pedersen, K. 1997. Microbial life in deep granitic rock. *FEMS Microbiology Reviews* 20:399–414. doi: 10.1111/j.1574-6976.1997.tb00325.x.

Pehme, P.E., J.P. Greenhouse, and B.L. Parker. 2007. The active line source temperature logging technique and its application in fractured rock hydrogeology. *Journal of Environmental and Engineering Geophysics* 12(4):307–322.

Pehme, P.E., B.L. Parker, J.A. Cherry, and J.P. Greenhouse. 2010. Improved resolution of ambient flow through fractured rock with temperature logs. *Groundwater* 48(2):191–205.

Pennell, K.D., G.A. Pope, and L.M. Abriola. 1996. Influence of viscous and buoyancy forces on the mobilization of residual tetrachloroethylene during surfactant flushing. *Environmental Science & Technology* 30(4):1328–1335.

Persoff, P., and K. Pruess. 1995. Two-phase flow visualization and relative permeability measurement in natural rough-walled rock fractures. *Water Resources Research* 31(5):1175–1186.

Phadnis, H.S., and J.C. Santamarina. 2011. Bacteria in sediments: pore size effects. *Geotechnique Letters* 1(4):91–93.

Power, W.L., and T.E. Tullis. 1991. Euclidean and fractal models for the description of rock surface roughness. *Journal of Geophysical Research: Solid Earth* 96(B1):415–424.

Pratt, J.W., H. Raiffa, and R. Schlaifer. 1965. *Introduction to statistical decision theory*. New York: McGraw-Hill.

Price, R.M., Z. Top, J.D. Happell, and P.K. Swart. 2003. Use of tritium and helium to define groundwater flow conditions in Everglades National Park. *Water Resources Research* 39(9):1267. doi: 10.1029/2002WR001929.

Pruess, K. 1985. A practical method for modeling fluid and heat flow in fractured porous media. *Society of Petroleum Engineers Journal* 25(1). doi: 10.2118/10509-PA.

Pruess, K., and Y.W. Tsang. 1990. On two-phase relative permeability and capillary pressure of rough-walled rock fractures. *Water Resources Research* 26(9):1915–1926.

Puls, R.W., and M.J. Barcelona. 1996. *Low-flow (minimal drawdown) ground-water sampling procedures*. Washington, DC: Office of Research and Development, Office of Solid Waste and Emergency Response, U.S. Environmental Protection Agency.

Purkamo, L., M. Bomberg, M. Nyyssönen, I. Kukkonen, L. Ahonen, R. Kietäväinen, and M. Itävaara. 2013. Dissecting the deep biosphere: retrieving authentic microbial communities from packer-isolated deep crystalline bedrock fracture zones. *FEMS Microbial Ecology* 85:324–337.

Raiffa, H. 1968. *Decision analysis; Introductory lectures on choices under uncertainty*. Reading, MA: Addison-Wesley.

Ramirez, A.L., J.J. Nitao, W.G. Hanley, R. Aines, R.E. Glaser, S.K. Sengupta, K.M. Dyer, T.L. Hickling, and W.D. Daily. 2005. Stochastic inversion of electrical resistivity changes using a Markov Chain Monte Carlo approach. *Journal of Geophysical Research* 110:B02101. doi: 10.1029/2004JB003449.

Rangel-German, E., S. Akin, and L. Castanier. 2006. Multiphase-flow properties of fractured porous media. *Journal of Petroleum Science and Engineering* 51(3–4):197–213.

Read, T., O. Bour, V. Bense, T.L. Borgne, P. Goderniaux, M.V. Klepikova, R. Hochreutener, N. Lavenant, and V. Boschero. 2013. Characterizing groundwater flow and heat transport in fractured rock using fiber-optic distributed temperature sensing. *Geophysical Research Letters* 40(10):2055–2059.

Rebata-Landa, V., and J.C. Santamarina. 2006. Mechanical limits to microbial activity in deep sediments. *Geochemistry, Geophysics, Geosystems* 7(11).

Rebata-Landa, V., and J.C. Santamarina. 2011. Mechanical effects of biogenic nitrogen gas bubbles in soils. *Journal of Geotechnical and Geoenvironmental Engineering* 138(2):128–137.

Reed, M., and B. Hetland. 2002. *DREAM: A dose-related exposure assessment model technical description of physical-chemical fates components*. Paper presented at SPE International Conference on Health, Safety and Environment in Oil and Gas Exploration and Production, March 20–22, Kuala Lumpur, Malaysia.

Reimus, P.W., and T.J. Callahan. 2007. Matrix diffusion rates in fractured volcanic rocks at the Nevada test site: evidence for a dominant influence of effective fracture apertures. *Water Resources Research* 43(7).

Renshaw, C.E., J.S. Dadakis, and S.R. Brown. 2000. Measuring fracture apertures: A comparison of methods. *Geophysical Research Letters* 27(2):289–292.

Reynolds, D.A., and B.H. Kueper. 2002. Numerical examination of the factors controlling DNAPL migration through a single fracture. *Groundwater* 40(4):368–377.

Reynolds, D.A., and B.H. Kueper. 2003. Effective constitutive properties for dense nonaqueous phase liquid (DNAPL) migration in large fracture networks: A computational study. *Water Resources Research* 39(9).

Reynolds, D.A., E.H. Jones, M. Gillen, I. Yusoff, and D.G. Thomas. 2008. Electrokinetic migration of permanganate through low-permeability media. *Groundwater* 46(4):629–637.

Riaz, A., and H. Tchelepi. 2006. Influence of relative permeability on the stability characteristics of immiscible flow in porous media. *Transport in Porous Media* 64(3):315–338.

Richardson, J.G., J.K. Kerver, J.A. Hafford, and J.S. Osoba. 2013. Laboratory determination of relative permeability. *Journal of Petroleum Technology* 4(8):187–196.

Robinson, J., T. Johnson, and L. Slater. 2013a. Evaluation of known-boundary and resistivity constraints for improving cross-borehole DC electrical resistivity imaging of discrete fractures. *Geophysics* 78(3):D115–D127.

Robinson, J., L. Slater, T. Johnson, and A. Binley. 2013b. Strategies for characterization of fractured rock using cross-borehole electrical tomography. *The Leading Edge* 32(7):784–790. doi: 10.1190/tle320 70784.1.

Robinson, P. 1984. *Connectivity, flow and transport in network models of fractured media*. Ph.D. dissertation, Oxford University.

Rodríguez, D.J., and B.H. Kueper. 2013. Specification of matrix cleanup goals in fractured porous media. *Groundwater* 51(1):58–65.

Rogers, S., D. Elmo, G. Webb, and A. Catalan. 2015. Volumetric fracture intensity measurement for improved rock mass characterisation and fragmentation assessment in block caving operations. *Rock Mechanics and Rock Engineering* 48(2):633–649.

Ross, B., and N. Lu. 1999. Dynamics of DNAPL penetration into fractured porous media. *Groundwater* 37(1):140–147.

Ross, N., R. Villemur, L. Deschenes, and R. Samson. 2001. Clogging of a limestone fracture by stimulating groundwater microbes. *Water Research* 35(8):2029–2037.

Rubin, H., and R.W. Buddemeier. 1996. Transverse dispersion of contaminants in fractured permeable formations. *Journal of Hydrology* 176(1–4):133–151.

Rubin, Y., and J.J. Gómez-Hernández. 1990. A stochastic approach to the problem of upscaling of conductivity in disordered media: Theory and unconditional numerical simulations. *Water Resources Research* 26(4):691–701.

Rudnick, R. L., and D.M. Fountain. 1995. Nature and composition of the continental crust—a lower crustal perspective. *Reviews of Geophysics* 33(3):267–309.

Ruiz-Agudo, E., C.V. Putnis, and A. Putnis. 2014. Coupled dissolution and precipitation at mineral–fluid interfaces. *Chemical Geology* 383:132–146.

Ruland, W.W., J.A. Cherry, and S. Feenstra. 1991. The depth of fractures and active ground-water flow in a Clayey Till Plain in southwestern Ontario. *Groundwater* 29(3):405–417.

Russell, H.H., J.E. Mathews, and G.W. Sewell. 1992. TCE removal from contaminated soil and ground water. *EPA Ground Water Issue.* EPA/540/S-92/002. Washington, DC: U.S. Environmental Protection Agency. Available at http://www.epa.gov/superfund/remedytech/tsp/download/tce.pdf (accessed January 20, 2014).

Rutqvist, J. 2012. The geomechanics of CO_2 storage in deep sedimentary formations. *Geotechnical and Geological Engineering* 30(3):525–551.

Rutqvist, J., J. Noorishad, C.F. Tsang, and O. Stephansson. 1998. Determination of fracture storativity in hard rocks using high-pressure injection testing. *Water Resources Research* 34(10):2551–2560.

Sahimi, M. 1994. *Applications of percolation theory*. London, UK: Taylor & Francis.

Sakaki, T. 2005. *The role of fracture-matrix interaction in drying of unsaturated fractured rock*. Ph.D. thesis, University of Colorado.

Sandia National Laboratories. 2010. *DAKOTA: A multilevel parallel object-oriented framework for design optimization, parameter estimation, uncertainty quantification, and sensitivity analysis. Version 5.3.1, user's manual*. Albuquerque, NM: Sandia National Laboratories. Available at http://dakota.sandia.gov/docs/dakota/5.3.1/Users-5.3.1.pdf (accessed December 13, 2013).

Santamarina, J.C. 2002. *Soil behavior at the microscale: Particle forces*. Symposium on Soil Behavior and Soft Ground Construction Honoring Charles C. "Chuck" Ladd, October 5–6, 2001, Cambridge, MA.

Sarkar, A.K., and M.M. Sharma. 1990. Fines migration in two-phase flow. *Journal of Petroleum Technology* 42(5):646–652.

Savoie, J.G., and D.R. LeBlanc. 2012. *Comparison of no-purge and pumped sampling methods for monitoring concentrations of ordnance-related compounds in groundwater, Camp Edwards, Massachusetts Military Reservation, Cape Cod, Massachusetts, 2009–2010*. U.S. Geological Survey Scientific Investigations Report 2012-5084, 23 pp. Available at http://pubs.usgs.gov/sir/2012/5084 (accessed January 20, 2014).

Schaefer, C.E., R.M. Towne, S. Vainberg, J.E. McCray, and R.J. Steffan. 2010. Bioaugmentation for treatment of dense non-aqueous phase liquid in fractured sandstone blocks. *Environmental Science and Technology* 44(13):4958–4964.

Schaefer, C.E., J. McCray, K. Christensen, P. Altman, P. Clement, and J. Torlapati. 2011. *DNAPL dissolution in bedrock fractures and fracture networks*. SERDP Project ER-1554. P. 146.

Schaefer, C.E., R.M. Towne, D. Root, and J.E. McCray. 2012. Assessment of chemical oxidation for treatment of DNAPL in fractured sandstone blocks, *Journal of Environmental Engineering* 138(1):1–7.

Schulze-Makuch, D., and D.S. Cherkauer. 1998. Variations in hydraulic conductivity with scale of measurement during aquifer tests in heterogeneous, porous carbonate rocks. *Hydrogeology Journal* 6(2):204–215.

Schulze-Makuch, D., D.A. Carlson, D.S. Cherkauer, and P. Malik. 1999. Scale dependency of hydraulic conductivity in heterogeneous media. *Groundwater* 37(6):904–919.

Schwartz, F.W., and L. Smith. 1988. A continuum approach for modeling mass-transport in fractured media. *Water Resources Research* 24(8):1360–1372.

Segura, J.M., and I. Carol. 2008. Coupled HM analysis using zero-thickness interface elements with double nodes. Part I: Theoretical model. *International Journal for Numerical and Analytical Methods in Geomechanics* 32:2083–2101.

Selroos, J.-O., D.D. Walker, A. Ström, B. Gylling, and S. Follin. 2002. Comparison of alternative modeling approaches for groundwater flow in fractured rock. *Journal of Hydrology* 257:174–188.

Seol, S.J., J.H. Kim, S.J. Cho, and S.H. Chung. 2004. A radar survey at a granite quarry to delineate fractures and estimate fracture density. *Journal of Environmental and Engineering Geophysics* 9(2):53–62.

SERDP (Strategic Environmental Research and Development Program). 2001. *SERDP/ESTCP expert panel workshop on research and development needs for cleanup of chlorinated solvent sites*. Available at https://www.serdpestcp.org/content/download/8241/101242/version/1/file/WorkshopChlorSolvCleanup-2001.pdf (accessed January 20, 2014).

Shah, N.W., S.F. Thornton, S.H. Bottrell, and M.J. Spence. 2009. Biodegradation potential of MTBE in a fractured chalk aquifer under aerobic conditions in long-term uncontaminated and contaminated aquifer microcosms. *Journal of Contaminant Hydrology* 103(3):119–133.

Shapiro, A.M. 2002. Cautions and suggestions for geochemical sampling in fractured rock. *Ground Water Monitoring & Remediation* 22(3):151–164.

Shapiro, A.M. 2011. The challenge of interpreting environmental tracer concentrations in fractured rock and carbonate aquifers. *Hydrogeology Journal* 19:9–12.

Shapiro, A.M., and P.A. Hsieh. 1998. How good are estimates of transmissivity from slug tests in fractured rock? *Groundwater* 36:37–48. doi: 10.1111/j.1745-6584.1998.tb01063.x.

Shapiro, A.M., P.A. Hsieh, W.C. Burton, and G.J. Walsh. 2007. Integrated multi-scale characterization of ground-water flow and chemical transport in fractured crystalline rock at the Mirror Lake Site, New Hampshire. In *Subsurface Hydrology: Data Integration for Properties and Processes*, D.W.

Hyndman, F.D. Day-Lewis, and K. Singha, eds. Washington, DC: American Geophysical Union. Pp. 201–226.

Sharmeen, R., W.A. Illman, S.J. Berg, T.C.J. Yeh, Y.J. Park, E.A. Sudicky, and K. Ando. 2012. Transient hydraulic tomography in a fractured dolostone: Laboratory rock block experiments. *Water Resources Research* 48(10).

Shemeta, J., B. Goodway, M. Willis, and W. Heigl. 2012. Introduction to this special section: Passive seismic and microseismic-Part 2. *The Leading Edge* 31(12):1428–1435.

Shin, S.H., J. Choe, and J.M. Kang. 1999. Clogging phenomena in fractures by dissolved hydrocarbons. In *Environmental Modeling: Proceedings of the International Conference on Water, Environment, Ecology, Socioeconomics and Health Engineering (WEESHE): October 18–21, 1999, Seoul National University, South Korea.* Highlands Ranch, CO: Water Resources Publications.

Silver, B.J., T. Onstott, L.-H. Lin, G. Rose, C. Ralston, B. Sherwood-Lollar, S.M. Pfiffner, T.L. Kieft, and S. McCuddy. 2010. In situ cultivation of subsurface microorganisms in a deep mafic sill: Implications for SLiMEs. *Geomicrobiology Journal* 27:329–348.

Simmons, A.M., and L.A. Neymark. 2012. Conditions and processes affecting radionuclide transport. *The Geological Society of America Memoirs* 209:277–362.

Singha, K., and S.M. Gorelick. 2005. Saline tracer visualized with electrical resistivity tomography: Field scale moment analysis. *Water Resources Research* 41:W05023. doi: 05010.01029/02004WR003460.

Singha, K., and S.M. Gorelick. 2006. Hydrogeophysical tracking of 3D tracer migration: The concept and application of apparent petrophysical relations. *Water Resources Research* 42:W06422. doi: 06410. 01029/02005WR004568.

Singha, K., F.D. Day-Lewis, and J.W. Lane, Jr. 2007. Geoelectrical evidence of bicontinuum transport in groundwater. *Geophysical Research Letters* 34(12):L12401.

Singhal, B.B.S., and R.P. Gupta. 2010. *Applied hydrogeology of fractured rocks second edition*. Dordrecht, The Netherlands: Springer Netherlands.

Singurindy, O., and B. Berkowitz. 2005. The role of fractures on coupled dissolution and precipitation patterns in carbonate rocks. *Advances in Water Resources* 28(5):507–521.

Slater, L.D., D. Brown, and A. Binley. 1996. Determination of hydraulically conductive pathways in fractured limestone using cross-borehole electrical resistivity tomography. *European Journal of Environmental and Engineering Geophysics* 1:35–52.

Slater, L., M.D. Zaidman, A.M. Binley, and L.J. West. 1997a. Electrical imaging of saline tracer migration for the investigation of unsaturated zone transport mechanisms. *Hydrology and Earth System Sciences* 1(2):291–302.

Slater, L.D., A. Binley, and D. Brown. 1997b. Electrical imaging of fractures using ground-water salinity change. *Groundwater* 35(3):436–442.

Slough, K.J., E.A. Sudicky, and P.A. Forsyth. 1999. Numerical simulation of multiphase flow and phase partitioning in discretely fractured geologic media. *Journal of Contaminant Hydrology* 40(2):107–136.

Smith, R.T. 2010. *The effect of thermally and chemically enhanced biostimulation on the hydraulic properties of a discreet fracture network in a bedrock aquifer*. M.Sc. thesis, Department of Civil Engineering, Queen's University, Kingston, Ontario.

Snow, D.T. 1970. Frequency and apertures of fractures in rock. *International Journal of Rock Mechanics and Mining Sciences* 7(1):23–40.

Song, D.L., M.E. Conrad, K.S. Sorenson, and L. Alvarez-Cohen. 2002. Stable carbon isotopic fractionation during enhanced in-situ bioremediation of trichloroethylene. *Environmental Science and Technology* 36(10):2262–2268.

Sorek, R., and P. Cossart. 2010. Prokaryotic transcriptomics: A new view on regulation, physiology and pathogenicity. *Nature Reviews Genetics* 11:9–16.

Sousa, R.L., K. Karam, and H.H. Einstein. 2014. Exploration analysis for landslide risk management. *Georisk* 8(3):155–170.

Staël von Holstein, C.-A.S. 1974. *The concept of probability in psychological experiments*. Dordrecht, The Netherlands: Reidel.

Stasiewicz, P.H. 1981. *Improving the contractual aspects of underground construction*. Engineering thesis. Massachusetts Institute of Technology.

Stearns, D.W. 1967. Certain aspects of fracture in naturally deformed rocks. In R.E. Riecker, *Rock mechanics seminar: [proceedings of] NSF Advanced Science Seminar in Rock Mechanics for College Teachers of Structural Geology, 26 June–28 July, 1967, Boston College*. Bedford, MA: Terrestrial Sciences Laboratory, Air Force Cambridge Research Laboratories.

Steefel, C.I., and P.C. Lichtner. 1994. Diffusion and reaction in rock matrix bordering a hyperalkaline fluid-filled fracture. *Geochimica et Cosmochimica Acta* 58(17):3595–3612.

Steimle, R. 2002. The state of the practice: Characterizing and remediating contaminated groundwater at fractured rock sites. *Remediation* 12:22–23.

Sterling, S.N. 1999. *Comparison of discrete depth sampling using rock core and a removable multilevel system in a TCE contaminated fractured sandstone*. Master's thesis, Department of Earth and Environmental Sciences, University of Waterloo, Waterloo, ON.

Sterling, S.N., B.L. Parker, J.A. Cherry, J.H. Williams, J.W. Lane, Jr., and F.P. Haeni. 2005. Vertical cross contamination of trichloroethylene in a borehole in fractured sandstone. *Groundwater* 43(4):557–573.

Stone, D. 1984. Sub-surface fracture maps predicted from borehole data: An example from the Eye-Dashwa Pluton, Atikokan, Canada. *International Journal of Rock Mechanics and Mining Sciences* 21(4):183–194.

Stoner, D.L., S.M. Watson, R.D. Stedtfeld, P. Meakin, L.K. Griffel, T.L. Tyler, L.M. Pegram, J.M. Barnes, and V.A. Deason. 2005. Application of stereolithographic custom models for studying the impact of biofilms and mineral precipitation on fluid flow. *Applied and Environmental Microbiology* 71(12):8721–8728.

Streltsova, T.D. 1988. *Well testing in heterogeneous formations*. New York: Wiley.

Sudicky, E.A., and E.O. Frind. 1982. Contaminant transport in fractured porous media—Analytical solutions for a system of parallel fractures. *Water Resources Research* 18(6):1634–1642.

Sudicky, E.A., and R.G. McLaren. 1992. The Laplace Transform Galerkin Technique for large-scale simulation of mass transport in discretely fractured porous formations. *Water Resources Research* 28(2):499–514.

Suman, A. 2013. *Joint inversion of production and time-lapse seismic data: Application to Norne Field.* Ph.D. thesis. Stanford University. Available at https://pangea.stanford.edu/departments/ere/dropbox/scrf/documents/Theses/SCRF-Theses/2010-2019/2013_PhD_Amit_Suman_Thesis.pdf (accessed January 20, 2014).

Swanson, R., K. Singha, F.D. Day-Lewis, A. Binley, K. Keating, and R. Haggerty. 2012. Direct geoelectrical evidence of mass transfer at the laboratory scale. *Water Resources Research* 48:W10543. doi: 10.1029/2012WR012431.

Szabo, Z., D.E. Rice, L.N. Plummer, E. Busenberg, S. Drenkard, and P. Schlosser. 1996. Age dating of shallow groundwater with chlorofluorocarbons, tritium/helium: 3, and flow path analysis, southern New Jersey coastal plain. *Water Resources Research* 32(4):1023–1038.

Szymczak, P., and A.J.C. Ladd. 2009. Wormhole formation in dissolving fractures. *Journal of Geophysical Research: Solid Earth* 114.

Szymczak, P., and A.J.C. Ladd. 2013. Interacting length scales in the reactive-infiltration instability. *Geophysical Research Letters* 40:3036–3041.

Takahashi, H.A., T. Nakamura, H. Tsukamoto, K. Kazahaya, H. Handa, and A. Hirota. 2013. Radiocarbon dating of groundwater in granite fractures in Abukuma Province, Northeast Japan. *Radiocarbon* 55(2):894–904.

Talley, J., G.S. Baker, M.W. Becker, and N. Beylre. 2005. Four dimensional mapping of tracer channelization in subhorizontal bedrock fractures using surface ground penetrating radar. *Geophysical Research Letters* 32:L04401. doi: 04410.01029/02004GL021974.

Tang, D.H., E.O. Frind, and E.A. Sudicky. 1981. Contaminant transport in fractured porous-media—analytical solution for a single fracture. *Water Resources Research* 17(3):555–564.

Tanguay, L.H., and G.M. Friedman. 2001. Petrophysical characteristics and facies of carbonate reservoirs: The Red River Formation (Ordovician), Williston basin. *American Association Petroleum Geologists Bulletin* 85(3):491–523.

Tatomir A.B., H. Class, R. Helmig, and A. Szymkiewicz. 2011. Modeling two phase flow in large scale fractured porous media with an extended multiple interacting continua method. *Computer Modeling in Engineering and Sciences* 77(2):81–111.

Taylor, S.W., and P.R. Jaffe. 1990a. Biofilm growth and the related changes in the physical properties of a porous medium. 1. Experimental investigation. *Water Resources Research* 26(9):2153–2159.

Taylor, S.W., and P.R. Jaffe. 1990b. Biofilm growth and the related changes in the physical properties of a porous medium. 3. Dispersivity and model verification. *Water Resources Research* 26(9):2171–2180.

Taylor, S.W., P.C.D. Milly, and P.R. Jaffe. 1990. Biofilm growth and the related changes in the physical properties of a porous medium: 2. Permeability. *Water Resources Research* 26(9):2161–2169.

Tentschert, E., A. Vigl, and A. Goricki. 2003. Guideline for geomechanical design for underground structures with continuous excavation. *Felsbau* 21(4):20–25.

Terzaghi, K. 1961. Past and future of applied soil mechanics. *Journal of the Boston Society of Civil Engineers* 48(2):110–139.

Tiedeman, C. R., P.J. Lacombe, and D.J. Goode. 2010. Multiple well-shutdown tests and site-scale flow simulation in fractured rocks. *Groundwater* 48:401–415. doi: 10.1111/j.1745-6584.2009.00651.x.

Tonkin, M., and J. Doherty. 2009. W00B10—Calibration-constrained Monte Carlo analysis of highly parameterized models using subspace techniques. *Water Resources Research* 45(12). doi: 10.1029/2007WR006678.

Topp, G.C., J.L. Davis, and A.P. Annan. 1980. Electromagnetic determination of soil water content: Measurements in coaxial transmission lines. *Water Resources Research* 16(3):574–582.

Treadgold, G., C. Sicking, V. Sublette, and G. Hoover. 2008. Azimuthal processing for fracture prediction and image improvement. *SEG Technical Program Expanded Abstracts* 27(1):988–992.

Triplett Kingston, J.L. 2008. *A critical evaluation of in-situ thermal technologies*. Ph.D. thesis, Arizona State University.

Triplett Kingston, J.L., P.R. Dahlen, and P.C. Johnson. 2010. State-of-the-practice review of in situ thermal technologies. *Ground Water Monitoring & Remediation* 30(4):64–72.

Tsang, C.F., and I. Neretnieks. 1998. Flow channeling in heterogeneous fractured rocks. *Reviews of Geophysics* 36(2):275–298.

Tsang, Y.W. 1984. The effect of tortuosity on fluid flow through a single fracture. *Water Resources Research* 20(9):1209–1215.

Tsang, Y.W., and C.F. Tsang. 1989. Flow channeling in a single fracture as a two-dimensional strongly heterogeneous permeable medium. *Water Resources Research* 25(9):2076–2080.

Tsoflias, G.P., and M.W. Becker. 2008. Ground-penetrating-radar response to fracture-fluid salinity: Why lower frequencies are favorable for resolving salinity changes. *Geophysics* 73(5):J25–J30. doi: 10.1190/1191.2957893.

Tsoflias, G.P., J.P. Van Gestel, P.L. Stoffa, D.D. Blankenship, and M.K. Sen. 2004. Vertical fracture detection by exploiting the polarization properties of ground-penetrating radar signals. *Geophysics* 69(3):803–810.

Tunnicliffe, B.S., and N.R. Thomson. 2004. Mass removal of chlorinated ethenes from rough-walled fractures using permanganate. *Journal of Contaminant Hydrology* 75(1–2):91–114.

Tweed, S.O., T.R. Weaver, and I. Cartwright. 2005. Distinguishing groundwater flow paths in different fractured-rock aquifers using groundwater chemistry: Dandenong Ranges, southeast Australia. *Hydrogeology Journal* 13(5–6):771–786.

Ulusay, R., and J.A. Hudson. 2007. *The Complete ISRM Suggested Methods for Rock Characterization, Testing and Monitoring: 1974–2006*. Ankara, Turkey: Commission on Testing Methods, International Society of Rock Mechanics.

Used Fuel Disposition Campaign. 2012. *Used Fuel Disposition Campaign disposal research and development roadmap: FCR&D-SED-2011-000065 REV 1*. Available at http://energy.gov/sites/prod/files/2013/08/f2/UFDDisposalRDRoadmapR1.pdf (accessed March 31, 2014).

Valdes, J.R., and J.C. Santamarina. 2006. Particle clogging in radial flow: Microscale mechanisms. *SPE Journal* 11(2):193–198.

VanderKwaak, J.E., and E.A. Sudicky. 1996. Dissolution of non-aqueous-phase liquids and aqueous-phase contaminant transport in discretely-fractured porous media. *Journal of Contaminant Hydrology* 23(1–2):45–68.

Venter, J.C., K. Remington, J.F. Heidelberg, A.L. Halpern, D. Rusch, J.A. Eisen, D. Wu, I. Paulsen, K.E. Nelson, W. Nelson, D.E. Fouts, S. Levy, A.H. Knap, M.W. Lomas, K. Nealson, O. White, J. Peterson, J. Hoffman, R. Parsons, H. Baden-Tillson, C. Pfannkoch, Y.-H. Rogers, and H.O. Smith. 2004. Environmental genome shotgun sequencing of the Sargasso Sea. *Science* 304:66–74.

Verges, E., D. Tost, D. Ayala, E. Ramos, and S. Grau. 2011. 3D pore analysis of sedimentary rocks. *Sedimentary Geology* 234(1–4):109–115.

Vesselinov, V.V., and D.R. Harp. 2012. *Model Analysis and Decision Support (MADS) for complex physical models*. XIX International Conference on Water Resources. University of Illinois at Urbana-Champaign, June 17–22. Available at http://mads.lanl.gov/papers/mads%20cmwr%2020120312.pdf (accessed December 18, 2013).

Vesselinov, V.V., S.P. Neuman, and W.A. Illman. 2001. Three-dimensional numerical inversion of pneumatic cross-hole tests in unsaturated fractured tuff: 1. Methodology and borehole effects. *Water Resources Research* 37(12):3001–3017.

Vieites, J.M., M.-E. Guazzaroni, A. Beloqui, P.N. Golyshin, and M. Ferrer. 2009. Metagenomics approaches in systems microbiology. *FEMS Microbiology Reviews* 33:236–255.

Vilks, P., L.H. Frost, D.B. Bachinski. 1997. Field-scale colloid migration experiments in a granite fracture. *Journal of Contaminant Hydrology* 26(1–4):203–214.

Vozoff, K., and D.L.B. Jupp. 1975. Joint inversion of geophysical data. *Geophysical Journal International* 42(3):977–991.

Vroblesky, D.A., and W.T. Hyde. 1997. Diffusion samplers as an inexpensive approach to monitoring VOCs in ground water. *Ground Water Monitoring & Remediation* 17(3):177–184.

Wagner, D., A.E. Milodowski, J.M. West, J. Wragg, and H. Yoshikawa. 2013. Mineralogical comparisons of experimental results investigating the biological impacts on rock transport processes. *Environmental Science-Processes & Impacts* 15(8):1501–1510.

Wang, H., C. Forster, and M. Deo. 2008. *Simulating naturally fractured reservoirs: Comparing discrete fracture network models to the upscaled equivalents*. April 20–23. Presentation at the AAPG Annual Convention, San Antonio, TX.

Wang, L.Y., S.C. Cowin, S. Weinbaum, and S.P. Fritton. 2000. Modeling tracer transport in an osteon under cyclic loading. *Annals of Biomedical Engineering* 28(10):1200–1209.

Warpinski, N.R., and L.W. Teufel. 1987. Influence of geologic discontinuities on hydraulic fracture propagation. *Journal of Petroleum Technology* 39(02):209–220.

Warpinski, N.R., M. Mayerhofer, K. Agarwal, and J. Du. 2013. Hydraulic-fracture geomechanics and microseismic-source mechanisms. *SPE Journal* 18(04):766–780.

Watanabe, K., Y. Sugumura, Y. Morita, and T. Tatsuya. 1997. Channel network modeling of the fractured granite in the Hinachi area, Japan. *Developments in Geotechnical Engineering* 81:247–262. Available at http://www.sciencedirect.com/science/article/pii/S0165125097800168 (accessed April 1, 2014).

Weisbrod, N., O. Dahan, and E.M. Adar. 2002. Particle transport in unsaturated fractured chalk under arid conditions. *Journal of Contaminant Hydrology* 56(1–2):117–136.

Werner, P.G., and M.F. Helmke. 2003. Chemical oxidation of tetrachloroethene in a fractured saprolite/bedrock aquifer. *Remediation Journal* 14(1):95–107.

Wiedermeier, T.H., M.A. Swanson, D.E. Moutoux, E.K. Gordon, J.T. Wilson, B.H. Wilson, D.H. Kampbell, P.E. Haas, R. N. Miller, J.E. Hansen, and F.H. Chapelle. 1998. *Technical protocol for evaluating natural attenuation of chlorinated solvents in ground water*. EPA/600/R-98/128.

Cincinnati, OH: National Risk Management Research Laboratory, Office of Research and Development, U.S. Environmental Protection Agency.

Will, R., R. Archer, and B. Dershowitz. 2005. Integration of seismic anisotropy and reservoir-performance data for characterization of naturally fractured reservoirs using discrete-feature-network models. *SPE Reservoir Evaluation and Engineering* 8(2):132–142.

Wilson, E.B.J., J.C. Decius, and P.C. Cross. 1980. *Molecular vibrations: The theory of infrared and Raman vibrational spectra.* New York: Dover.

Winberg, A., ed. 2010. *Fault rock zones characterization—Final report, TRUE-1 continuation project*. TR-10-36. Stockholm, Sweden: Svensk kärnbränslehantering.

Winberg, A., P. Andersson, J. Byegård, A. Poteri, V. Cvetkovic, W. Dershowitz, T. Doe, J. Hermanson, J. J. Gómez-Hernández, A. Hautojärvi, D. Billauz, E.-L. Tullborg, D. Holton, P. Meier, and A. Medina. 2003. *Final report of the TRUE Block Scale Project: 4. Synthesis of flow, transport and retention in the block scale*. SKB Technical Report TR-02-16. Available at www.skb.se/upload/publications/pdf/TR-02-16.pdf (accessed December 6, 2013).

Würdermann, H., F. Möller, M. Kühn, W. Heidug, N.P. Christensen, G. Borm, and F.R. Schilling. 2010. CO2SINK-From site characterization and risk assessment to monitoring and verification: One year of operational experience with the field laboratory for CO_2 storage at Ketzin, Germany. *International Journal of Greenhouse Gas Control* 4(6):938–951.

Xu, C., and P. Dowd. 2010. A new computer code for discrete fracture network modelling. *Computers and Geosciences* 36(3):292–301.

Yamano, M., and S. Goto. 2005. Long-term monitoring of the temperature profile in a deep borehole: Temperature variations associated with water injection experiments and natural groundwater discharge. *Physics of the Earth and Planetary Interiors* 152(4):326–334.

Yang, Y.R., and P.L. McCarty. 2000. Biologically enhanced dissolution of tetrachloroethene DNAPL. *Environmental Science & Technology* 34(14):2979–2984.

Yasuhara, H., A. Polak, Y. Mitani, A.S. Grader, P.M. Halleck, and D. Elsworth. 2006. Evolution of fracture permeability through fluid–rock reaction under hydrothermal conditions. *Earth and Planetary Science Letters* 244(1–2):186–200.

Yeh, T.C.J., and J. Simunek. 2002. Stochastic fusion of information for characterizing and monitoring the vadose zone. *Vadose Zone Journal* 1:207–221.

Yeh, W.G. 1986. Review of parameter identification procedures in groundwater hydrology: The inverse problem. *Water Resources Research* 22(2):95–108.

You, Z., A. Badalyan, P. Bedrikovetsky, M. Hand, and D. Jenson. 2013. *Productivity decline due to fines migration (modelling and field case)*. International Petroleum Technology Conference, March 26–28, Beijing, China. doi: 10.2523/17059-MS.

Yow, J.L., and J.R. Hunt. 2002. Coupled processes in rock mass performance with emphasis on nuclear waste isolation. *International Journal of Rock Mechanics and Mining Sciences* 39:143–150.

Yu, J., and K. Pinder. 1994. Effective diffusivities of volatile fatty acids in methanogenic biofilms. *Bioresource Technology* 48(2):155–161.

Yun, T.S., B. Dumas, and J.C. Santamarina. 2011. Heat transport in granular materials during cyclic fluid flow. *Granular Matter* 13(1):29–37.

Zafirovski, Z., I. Pesevski, and J. Papic. 2012. Methodology for extrapolation of rock mass deformability parameters in tunneling. *Facta Universitatis—Series: Architecture and Civil Engineering* 10(3):235–244.

Zaidman, M., R. Middleton, L. West, and A. Binley. 1999. Geophysical investigation of unsaturated zone transport in the Chalk in Yorkshire. *Quarterly Journal of Engineering Geology* 32:185–198.

Zha, Y., T.C.J. Yeh, D. Mao, J. Yang, and W. Lu. 2014. Usefulness of flux measurements during hydraulic tomographic survey for mapping hydraulic conductivity distribution in a fractured medium. *Advances in Water Resources* 71:162–176.

Zhang, L., and X. Ding. 2010. Variance of non-parametric rock fracture mean trace length estimator. *International Journal of Rock Mechanics and Mining Sciences* 47(7):1222–1228. https://doi.org/10.1016/j.ijrmms.2010.06.014.

Zhang, L.M., and D.G. Fredlund. 2003. Characteristics of water retention curves for an unsaturated fractured rock mass. *Proceedings of the Second Asian Conference on Unsaturated Soils.* UNSAT-ASIA. April 15–17. Osaka, Japan. Pp. 425–429.

Zhang, W., X. Tang, N. Weisbrod, and Z. Guan. 2012. A review of colloid transport in fractured rocks. *Journal of Mountain Science* 9(6):770–787.

Zhong, L., J. Szecsody, M. Oostrom, M. Truex, X. Shen, and X. Li. 2011. Enhanced remedial amendment delivery to subsurface using shear thinning fluid and aqueous foam. *Journal of Hazardous Materials* 191(1-3):249–257.

Zimmermann, G., G. Blöcher, A. Reinicke, and W. Brandt. 2011. Rock specific hydraulic fracturing and matrix acidizing to enhance a geothermal system—Concepts and field results. *Tectonnophysics* 503(1–2):146–154.

Zoback, M.D. 2007. *Reservoir Geomechanics*. Cambridge, UK: Cambridge University Press.

Zoback, M.D. 2010. *Reservoir Geomechanics.* Cambridge, UK: Cambridge University Press.

Zoback, M.D., F. Rummel, R. Jung, and C.B. Raleigh. 1977. Laboratory hydraulic fracturing experiments in intact and pre-fractured rock. *International Journal of Rock Mechanics and Mining Sciences* 14(2):49–58.

Zuber, A., and J. Motyka. 1994. Matrix porosity as the most important parameter of fissured rocks for solute transport at large scales. *Journal of Hydrology* 158(1–2):19–46.

Zytner, R.G. 1994. Sorption of benzene, toluene, ethylbenzene and xylenes to various media. *Journal of Hazardous Materials* 38(1):113–126.

Zyvoloski, G., E. Kwicklis, A.A. Eddebbarh, B. Arnold, C. Faunt, B.A. Robinson. 2003. The site-scale saturated zone flow model for Yucca Mountain: Calibration of different conceptual models and their impact on flow paths. *Journal of Contaminant Hydrology* 62–63:731–750.

Appendix A

Committee Member Biographies[1]

David E. Daniel (NAE) (*Chair*), has served as the president of The University of Texas at Dallas since 2005. From 1996 to 2005, he served in multiple roles, finishing as the dean of engineering at the University of Illinois. He served on the faculty at The University of Texas at Austin from 1980 to 1996. Dr. Daniel has been recognized for his leadership in waste containment, landfilling of wastes, and cleanup of contaminated lands. He has worked on the flow of water and chemicals in soils, the engineering design of soil barriers (e.g., clay liners) and drainage systems for waste containment systems, the measurement of hydraulic conductivity in the laboratory and field, the alterations of barrier materials caused by chemicals, the construction of waste containment systems, and various design and permitting issues. The work has focused on bottom liner systems for landfills, final cover systems for landfills and abandoned dumps, the containment of buried wastes or contaminated groundwater, and the cleanup of old waste disposal sites. He has also conducted research on various types of geosynthetic materials, with most of the work involving geosynthetic clay liners used for waste containment but some of the work involving geomembranes, geonets, and geotextiles. Dr. Daniel's professional work has been recognized by the American Society of Civil Engineers, which awarded him its highest honor for papers published in its journals (the Norman Medal), and on two separate occasions awarded him its second highest honor (the Croes Medal). He received the Presidents' Award in 2007 and the Outstanding Projects and Leaders Award for Education in 2010. Dr. Daniel received his bachelor's degree, master's degree, and Ph.D. in civil engineering from The University of Texas at Austin.

Lisa Alvarez-Cohen (NAE), is the Fred and Claire Sauer Professor of Environmental Engineering in the Department of Civil and Environmental Engineering at the University of California, Berkeley. Her research areas include environmental microbiology and ecology, biotransformation and the fate of environmental contaminants, nutrient cycling in soils, and innovative molecular and isotopic techniques for studying the microbial ecology of complex communities. Specifically, her research focuses on the application of omics-based molecular tools and isotopic techniques to understand and optimize microbial communities involved in the bioremediation of emerging and conventional environmental contaminants and nutrient cycling in engineered processes. Bioremediation and nutrient cycling are processes that rely on complex mixed microbial communities that interact to catalyze important reaction pathways. Dr. Alvarez-Cohen is a fellow of the American Academy of Microbiology, an editorial advisory board member of *Environmental Science and Technology*, and an associate editor of *Environmental Engineering Science*. Her previous service to the National Academies of Sciences, Engineering, and Medicine includes numerous committees, including the Water Science and Technology Board, Committee on Metagenomics: Challenges and Functional Applications, Committee to Assess the Performance of Engineered Barriers, Committee on In Situ Bioremediation, and Committee on Source Removal of Contaminants in the Subsurface. She received her B.S. in engineering and applied science from Harvard University and her M.S. and Ph.D. in environmental engineering and science from Stanford University.

[1] Biography information current as of September 2015.

William Dershowitz is an engineer and hydrogeologist at Golder Associates Inc., in Redmond, Washington. Dr. Dershowitz has a broad background in the analysis and modeling of fractured rock. In addition to more than 35 years of experience in conventional hydrogeologic techniques and modeling, Dr. Dershowitz is a pioneer of the Discrete Fracture Network (DFN) approach to flow, transport, and geomechanics. Since 1977, Dr. Dershowitz has developed and applied DFN models for environmental, civil, mining, geothermal, and oil/gas projects. Dr. Dershowitz integrates the principles of geology, structural geology, geophysics, hydrodynamics, and geomechanics to develop models for groundwater flow and for transport pathways and retention. He is also active in the development of approaches for hydrogeological optimization and uncertainty analysis for fractured and heterogeneous aquifers and is the author of more than 50 professional papers. Dr. Dershowitz earned his B.S., M.S., and Ph.D. in civil engineering (geotechnics/rock mechanics) from the Massachusetts Institute of Technology. He has served on the Board of Directors of the American Rock Mechanics Association and holds an adjunct faculty appointment at the University of Washington.

Herbert H. Einstein is a professor in the Civil and Environmental Engineering Department at the Massachusetts Institute of Technology. Dr. Einstein is a former chair of the U.S. National Committee on Rock Mechanics and is well known in the civil engineering community for his work in rock mechanics. His areas of expertise include rock fracture genesis, fracture coalescence, description of fracture patterns, and hydrologic properties of rock masses. He is particularly well known for his work on fracture pattern characterization, including stochastic representation of fracture patterns, flow in individual fractures, and fractured rock masses and hydraulic fracturing. His research and consulting activities have included the influence of fractured rock patterns on the performance of nuclear waste storage facilities and engineered geothermal systems. Dr. Einstein earned his Dipl.-Bauing and Sc.D. from Eidgenössische Technische Hochschule in Zurich.

Carl Gable is the group leader of the Computational Earth Science Group in the Earth and Environmental Sciences Division at the Los Alamos National Laboratory. His major research interests include two- and three-dimensional unstructured finite element mesh generation and model setup for geological applications, flow and reactive chemical transport modeling in saturated and unsaturated porous media, computational physics and fluid dynamics, and the interaction of tectonic plates in mantle convection. Dr. Gable and collaborators have developed capabilities to build high-quality computational meshes of large Discrete Fracture Networks that are optimized for parallel multi-phase, multi-component flow solvers and transport using Lagrangian particle tracking. The areas of subsurface flow and transport modeling applications in which he works are underground waste repositories, unconventional fossil energy, geothermal energy, carbon capture, storage and utilization, water resource management, and the remediation of contaminated groundwater. Dr. Gable earned his B.A. in geophysics from the University of California, Berkeley, and his M.S. and Ph.D. in applied physics and geophysics from Harvard University.

Franklin M. Orr, Jr. (resigned from the committee December 2014) (NAE), was sworn in as the Under Secretary for Science and Energy on December 17, 2014. As the Under Secretary, Dr. Orr is the principal advisor to the Secretary and Deputy Secretary on clean energy technologies and science and energy research initiatives. Prior to joining the U.S. Department of Energy, he was the Keleen & Carlton Beal Professor in the Department of Energy Resource Engineering at Stanford University. He joined Stanford in 1985. He served as the founding director of the Precourt

Institute for Energy at Stanford University from 2009 to 2013. He was the founding director of the Stanford Global Climate & Energy Project from 2002 to 2008, and he served as the dean of the School of Earth, Energy & Environmental Sciences at Stanford from 1994 to 2002. He was head of the miscible flooding section at the Petroleum Recovery Research Center, New Mexico Institute of Mining and Technology, from 1978 to 1985; a research engineer at the Shell Development Company Bellaire Research Center from 1976 to 1978; and an assistant to the director, Office of Federal Activities, U.S. Environmental Protection Agency, from 1970 to 1972. He holds a Ph.D. from the University of Minnesota and a B.S. from Stanford University, both in chemical engineering.

David Reynolds is an associate at Geosyntec Consultants. Dr. Reynolds's primary areas of expertise include hazardous waste management with a particular focus on groundwater remediation, fate and transport of chemical contaminants in the environment, and site investigation in fractured systems. He has been the technical director, reviewer, or expert witness on numerous site investigation and remediation projects in fractured and unfractured systems during his 20 years in the industry. Dr. Reynolds was a faculty member and leader of the Hydrogeology Research Group at The University of Western Australia and the research director of the National Centre for Groundwater Research and Training, Flinders University, Adelaide, Australia. His research interests include the migration of contaminants in fractured consolidated and unconsolidated media, remediation of low permeability soils and rock, and value of information approaches in site investigation. Dr. Reynolds received a B.A.Sc. in geological engineering from the University of Waterloo, an M.Sc. (Eng.) from Queen's University, and a Ph.D. in environmental engineering from Queen's University.

J. Carlos Santamarina (NAE) is a professor at the King Abdullah University of Science and Technology, Saudi Arabia, formerly at the Georgia Institute of Technology in the United States. His research focuses on the fundamental study of geomaterials and subsurface coupled processes at multiple scales. The implementation of this research has involved the development and utilization of multi-scale experimental methods, high-resolution process monitoring, forward modeling, and inverse problem solving. Using this theoretical and experimental research framework, Dr. Santamarina and co-workers explore critical problems in energy geoengineering and science, with emphasis on petroleum, gas hydrates, and carbon dioxide geological storage. He has co-authored 2 books and more than 300 articles that summarize salient concepts and research results. He is a corresponding member of the National Academy of Sciences of Argentina and the U.S. National Academy of Engineering. He holds a Ph.D. from Purdue University, an M.S. from the University of Maryland, and a B.Sc. from Universidad de Córdoba.

Allen Shapiro is a research hydrologist with the U.S. Geological Survey in Reston, Virginia. His research has focused on characterizing groundwater flow and chemical transport in fractured rock. It has included investigations in various geologic settings, including fractured and dissolution-enhanced limestone, bedded sedimentary formations, and igneous and metamorphic rock. Dr. Shapiro has authored papers on equipment design and field techniques, the interpretation of hydraulic and geochemical data, and theories of groundwater flow and chemical transport. His research has application to issues of societal importance, including water supply, groundwater contamination and restoration, waste isolation, and groundwater flow in the vicinity of engineered structures. Dr. Shapiro earned a bachelor's degree in civil engineering from Lafayette College in

Easton, Pennsylvania, and a master's and a Ph.D. in civil and geological engineering from Princeton University.

Kamini Singha is an associate professor in the Department of Geology and Geological Engineering and the associate director of the Hydrologic Science and Engineering Program at the Colorado School of Mines. She worked at the U.S. Geological Survey Hydrogeophysics Branch from 1997 to 2000 and served on the faculty of The Pennsylvania State University from 2005 to 2012. Her research interests are focused on the physical process controlling solute and contaminant mass transport, including "long-tailed" distributions of solute arrival times in groundwater systems and during groundwater–surface water exchange, integration of geophysical imaging with flow and transport modeling, and establishing field-scale rock physics relations between geophysical and hydrogeologic parameters. Dr. Singha served as the chair of the American Geophysical Union's Hydrogeophysics Technical Committee from 2009 to 2012 and is an associate editor at *Water Resources Research*. She earned her B.S. in geophysics from the University of Connecticut and her Ph.D. in hydrogeology from Stanford University.

Appendix B

Open Session Agendas

MEETING 1—January 7, 2013

9:00 a.m.	**Welcome and introductions** *David Daniel (NAE), Ph.D., Chair*, Committee on Characterization, Modeling, Monitoring, and Remediation of Fractured Rocks
9:05	**Sponsor expectations** *Thomas Nicholson*, U.S. Nuclear Regulatory Commission, Division of Risk Analysis, Office of Nuclear Regulatory Research *Mark Schoppet*, National Aeronautics and Space Administration, Environmental Management Division
10:50	**Break**
11:00	**Perspectives on fractured bedrock across disciplines and scales** *Matthew Becker, Ph.D.*, California State University, Long Beach
12:00 p.m.	**Working lunch in meeting room; continued discussions**
12:30	**A recent study of fault zone hydrology** *Kenzi Karasaki, Ph.D.*, Lawrence Berkeley National Laboratory
1:30	**Sponsor expectations** (cont.) *TBD*, U.S. Department of Energy, Office of Nuclear Fuel Disposition
2:00	**Open session adjourns**

WORKSHOP—May 29–31, 2013

DAY ONE: Wednesday, May 29, 2013

CHARACTERIZING PHYSICAL AND BIOGEOCHEMICAL PROCESSES IN FRACTURED ROCK

8:00 a.m.	**Continental breakfast**
8:30	**Welcome and introductory remarks** *David Daniel (NAE), Ph.D., Committee Chair*

SESSION 1: RECENT ADVANCES IN UNDERSTANDING GEOLOGIC AND CHEMICAL PROCESSES

Session Moderator: Allen Shapiro, Ph.D., committee member

8:50	**Fractured rock: How did it get that way?** *Terry Engelder, Ph.D.*, The Pennsylvania State University
9:20	**Hydromechanical coupling** *Mark Zoback (NAE), Ph.D.*, Stanford University
9:50	**Contaminant geochemical behavior in fractured rocks: Lessons from porous media** *Richelle Allen-King, Ph.D.*, University at Buffalo
10:20	**Break**
10:40	**Panel discussion**
11:40	**Lunch**

SESSION 2: ENHANCED CHARACTERIZATION

Session Moderator: Kamini Singha, Ph.D., committee member

1:00 p.m.	**Fracture architecture and mechanical stratigraphy in sedimentary rocks** *Michael Gross, Ph.D.*, Shell Oil
1:30	**Advances in the characterization of fractured rock systems beyond the borehole wall using geophysical imaging** *Lee Slater, Ph.D.*, Rutgers University
2:00	**Panel discussion**
3:00	**Break**

SESSION 3: SYNTHESIZING CHARACTERIZATION INFORMATION INTO CONCEPTUAL MODELS OF GROUNDWATER FLOW AND BIOGEOCHEMICAL TRANSPORT

Session Moderator: Bill Dershowitz, Ph.D., committee member

3:20	**Experiences from site-descriptive and safety assessment modeling, and input from ongoing research and development plans** *Jan-Olof Selroos, Ph.D.*, Swedish Nuclear Fuel and Waste Management Company (SKB)
3:50	**Data model synthesis for flow and transport in fractured rocks** *Paul Hsieh, Ph.D.*, U.S. Geological Survey

4:20 **Panel discussion**
Jan-Olof Selroos, Ph.D., SKB
Paul Hsieh, Ph.D., U.S. Geological Survey
Kent Novakowski, Ph.D., Queens University
Scott Painter, Ph.D.

5:20 **Adjourn for the day**

DAY TWO: Thursday, May 30, 2013

REMEDIATION AND MONITORING IN FRACTURED ROCK

8:00 a.m. **Continental breakfast**

8:30 **Welcome and introductory remarks**
David Daniel (NAE), Ph.D., Committee Chair

SESSION 4: REMEDIATION IN FRACTURED ROCK

Session Moderator: David Reynolds, Ph.D., committee member

8:45 **What is the future of remediation science and technology and how do we prepare?**
Jeffrey Marqusee, Ph.D., Strategic Environmental Research and Development Program and Environmental Security Technology Certification Program

9:15 **Use of thermal conductive heating to remove chlorinated solvents from fractured bedrock**
Bernie Kueper, Ph.D., Queens University

9:45 **Biotic/abiotic degradation**
Leo Lehmicke, Ph.D., CO2 and Water, Inc.

10:15 **Break**

10:30 **In situ bioremediation of uranium**
Steve Yabusaki, Pacific Northwest National Laboratory

11:00 **Panel discussion**

12:00 p.m. **Lunch**

SESSION 5: MONITORING FOR EARLY INDICATORS OF ENGINEERING PERFORMANCE AND REMEDIATION PROGRESS

Session Moderator: Lynn Orr, Jr., Ph.D., committee member

1:00 **Session introduction**

1:15 **Working group discussions**

Questions for working groups:

1. What processes could be monitored as early indicators of good or bad performance of engineered systems?
2. Given monitoring data collection technologies, what strategies can be applied to get more out of collected data?
3. What new approaches to monitoring could be applied, given current technologies, that might reduce performance uncertainties in monitoring for radionuclides, organics, and metals?
4. What new and emerging technologies employing new strategic approaches could be applied to monitor system performance and reduce uncertainties in monitoring for radionuclides, organics, and metals? What do we do to get there?

2:30 **Working group summaries provided in plenary session**

3:00 **Large group discussion**

CONCLUDING REMARKS

4:00 **Closing remarks**
David Daniel (NAE), Ph.D., Committee Chair

4:30 **Workshop adjourns**